T0220469

Physics Around Us

How and Why Things Work

Ernest M Henley

J Gregory Dash

University of Washington, USA

Physics Around Us

How and Why Things Work

 **World Scientific**

NEW JERSEY · LONDON · SINGAPORE · BEIJING · SHANGHAI · HONG KONG · TAIPEI · CHENNAI

Published by

World Scientific Publishing Co. Pte. Ltd.

5 Toh Tuck Link, Singapore 596224

USA office: 27 Warren Street, Suite 401-402, Hackensack, NJ 07601

UK office: 57 Shelton Street, Covent Garden, London WC2H 9HE

British Library Cataloguing-in-Publication Data
A catalogue record for this book is available from the British Library.

PHYSICS AROUND US
How and Why Things Work

Copyright © 2012 by World Scientific Publishing Co. Pte. Ltd.

All rights reserved. This book, or parts thereof, may not be reproduced in any form or by any means, electronic or mechanical, including photocopying, recording or any information storage and retrieval system now known or to be invented, without written permission from the Publisher.

For photocopying of material in this volume, please pay a copying fee through the Copyright Clearance Center, Inc., 222 Rosewood Drive, Danvers, MA 01923, USA. In this case permission to photocopy is not required from the publisher.

ISBN-13 978-981-4350-63-1 (pbk)
ISBN-10 981-4350-63-X (pbk)

Printed in Singapore by World Scientific Printers.

This book is dedicated to
Elaine Henley and Joan Dash

Foreword

This book is intended for the serious student interested in Physics. The contents have been used for over a decade to teach students in the University of Washington's "Early Entrance Program."

These students, few of whom have previously been appropriately challenged, skip high school and take a strenuous year of "Transition School" before becoming freshmen at the University of Washington. They enter Transition School at about age 12–14 and generally graduate from the University at ages 18–19, many with double or triple majors. They tend to do very well at the University, and often win scholastic awards.

During the Transition School year, the students take English, Ethics, History, Mathematics, and Physics. The materials for this book have been used for Physics.

The book is suitable for beginning College or High School Physics courses. It does not use calculus, but assumes that the students know algebra. The book teaches the essentials of trigonometry and vector algebra that are required.

We want to thank all the students who have gone through the Transition School during the last twelve years for their questions and interest. The authors thank Oscar Vilches for correcting numerous errors in the text. We also thank Alvin Chong for his help with the editorial process.

Contents

Introduction

I. Scientific Notation

In physics it is very common to use scientific notation to avoid very large or very small numbers. For example, the radius of the earth is 6,380,000 m. In scientific notation this is 6.38×10^6 m. The Bohr radius for the hydrogen atom is 0.0000000000529 m; it becomes 5.29×10^{-10} m.

The scientific notation uses powers of 10: $10^0 = 1$, $10^1 = 10$, $10^2 = 100$, etc. Similarly $10^{-1} = 1/10 = 0.1$, $10^{-2} = 1/100 = 0.01$, etc. You will note that $10^{-n} = 1/10^n$, where n is any integer. This is useful when multiplying or dividing.

To multiply 10^n by 10^m you add the exponents; thus $10^n \times 10^m = 10^{n+m}$; similarly $10^n/10^m = 10^n \times 10^{-m} = 10^{n-m}$. $(10^n)^2 = 10^{2n}$. $\sqrt{10^n} = 10^{n/2}$.

If you want to add 10^n and 10^m, you must first make the exponents identical. For example $5 \times 10^2 + 5 \times 10^1 = 50 \times 10^1 + 5 \times 10^1 = 5 \times 10^2 + 0.5 \times 10^2 = 55 \times 10^1 = 5.5 \times 10^2$.

Use powers of 10 and you will soon become familiar with them.

Illustrative Example: Use scientific notation to add, subtract, multiply, and divide 0.0046 and 0.00004.

Answer: Multiplication: $4.6 \times 10^{-3} \times 4 \times 10^{-5} = 18.4 \times 10^{-8}$;
Division: $4.6 \times 10^{-3}/(4 \times 10^{-5}) = 1.15 \times 10^2$;
Addition, Subtraction: $4.6 \times 10^{-3} \pm 0.04 \times 10^{-3} = 4.64,\ 4.56 \times 10^{-3}$.

An average is denoted by a bar over the symbol; for example an average height equal to 2 meters can be written as $\bar{h} = 2$ m. Alternatively, an average can be denoted by a subscript, e.g., h_{av}.

The standard units in mechanics are lengths in meters, masses in kilograms (kg), and times in seconds (s), e.g., 10 m/s.

II. Significant Figures

Many students have trouble with significant figures. Significant figures are the number of digits that are meaningful or known with certainty. For instance a mass of 56 kg has 2 significant figures, whereas 56.1 or 56.0 kg has 3 significant figures. Similarly 5.0 (or 5.) has 2 significant figures. The added zero is meaningful, once it is given. A mass of 0.56 kg has two significant figures, as does the mass 0.00056 kg, but 0.560 kg has 3 significant figures. Here the scientific notation is particularly useful. The number $0.00056 = 5.6 \times 10^{-4}$. It is much easier to see that there are only 2 significant figures here.

The smallest number of significant figures in a problem is dominant. Thus, if you are given two masses of 56 kg and 56.1 kg, the sum is 110 kg because 56 kg only has 2 significant figures. Your hand held computer may give an answer to many more figures than are significant; for instance $56.1/56 = 1.0017857$, but the correct answer is 1.0 with 2 significant figures. When numbers are multiplied, divided, added, or subtracted, it is the smallest number of significant figures which holds sway.

When you are given two significant figures, you may have to "round up" or "round down" your answer. Thus for two significant figures, 12.4 becomes 12, 12.5 becomes 13 and 12.7 becomes 13.

In doing problems, answers should not be given to more significant figures than appear in the problem.

Illustrative Example: What do you get for the density if the mass is 2.756 kg and the volume is 25 cu cm?

Answer: 1.1×10^{-1} kg/cm^3

III. Dimensions or Units

You *must* specify units in your answers. The number 56 is meaningless by itself, but 56 m or 56 kg has a well defined meaning.

It is useful, furthermore, to carry dimensional units when you do a problem. The units that are not needed in the answer should cancel. An example is that the distance traveled in meters is the speed (in m/s) multiplied by the time in seconds. Thus 30 m/s × 2 s = 60 m. The seconds cancel out. On the other hand if the time is given as 2 minutes, you need to convert this time to seconds, 2 min × 60 s/min = 120 s, before proceeding.

You can catch mistakes, such as forgetting to change minutes to seconds, by keeping units at all times.

PART 1
Mechanics

Chapter 1

The Sizes of Typical Objects, and How We Know

1.1 Introduction

The standard way to learn physics is systematic, beginning with simple mechanics, and gradually going on to more exotic phenomena. We will follow that pattern, more or less. But to start, we will spend a little time with some of the advanced topics, as a taste of the physics to come.

We begin by exploring the sizes of physical objects. From atoms to galaxies, they range over a factor larger than 10^{30}, i.e. a one followed by 30 zeroes. Beyond the sheer impressiveness of such a large number is the remarkable fact that large and small physical objects obey many of the same physical laws, for example: Newtonian dynamics, gravitation and the conservation of energy. There are additional phenomena that become important at extreme conditions, e.g. quantum mechanics for very small things, and special relativity at very high speeds, but much of basic physics holds for atoms, baseballs and galaxies. In this and the following chapters we will explore how they all can move, interact, and follow the laws of *Classical Mechanics*.

1.2 Atoms

Democritus, in 4th century BC Greece, believed that matter is made of atoms, and Epicurus was inspired to write a poem about them. Here is an extract.

> *These atoms, which are separated from each other in the infinite void and distinguished from each other in shape, size and arrangement, move in the void, overtake each other, and collide*

> *.... Observe what happens when sunbeams are admitted into a*
> *building and shed light into shadowy places. You will see a mul-*
> *titude of tiny particles mingling in a multitude of ways in the*
> *empty space within the light of the beam, as though continuing*
> *in everlasting conflict, rushing into battle rank upon rank with*
> *never a moment's pause in a rapid sequence of unions and dis-*
> *unions. From this you may picture what it is for the atoms to*
> *be perpetually tossed about in the void.*

Democritus only imagined atoms, but we know now after many experiments that they do exist, and all of ordinary matter is made of them. Atoms and molecules, which are chemical clusters of atoms, are so tiny that the ultimate graininess of matter is undetectable to our unaided senses. The diameter of a small molecule is less than a *nanometer* (nm), 10^{-9} meters. Another convenient unit when dealing with atoms and molecules is the *Ångstrom*, abbreviated Å, which is equal to 0.1 nanometer.

We now have instruments to detect single atoms and molecules, but long before modern techniques were available, Agnes Pockels devised an experiment in her kitchen that gave an estimate of a molecule's size. (Universities in 19th century Germany did not admit female students. Women were allowed only to sit in the back of the lecture hall, so that "they would not disturb the men"). Barred from university attendance, Pockels invented experiments that she could do at home. When she described her observations in a letter to Lord Raleigh, one of England's scientific immortals, he was extremely impressed, and arranged for their publication. Here is a brief description of some of her work.

Pockels estimated the sizes of small molecules from experiments with surface films on water. Using a very sensitive surface pressure gauge of her own design, she measured the surface area of an oil film caused by the spreading of a small oil drop. She compressed the film until the surface pressure increased rapidly (now known as the *Pockels Point*); she surmised that at that point the molecules of the film were tightly packed and the film was only one molecule thick. Then from the diameter of the drop and the area of the film she estimated the size of a single molecule. You can repeat her experiment, as follows. Sprinkle a little pinch of flour over a large pan of water, so that the grains are distributed over the surface. Now form a very small drop of liquid soap on the tip of a pencil, and touch the tip to the water surface. Watch the drop spread, as it pushes the flour grains away and leaves a clear space. From the initial size of the drop and the cleared area, you can estimate the size of the molecules (see Problem 5).

1.3   Earth and the Solar System

The earth's spherical shape[a] was deduced long ago. Aristotle in his book
On the Heavens wrote

*If the Earth were not spherical, eclipses of the Moon would not exhibit seg-
ments of the shape that they do . . . they are always convex.* His explanation
was based on the realization that the Moon shines only by reflected light,
and the eclipses were due to the Earth's shadow.

And Strabo in the first century AD gave further proof: *. . . as sailors
approach dry land more and more of the shore becomes visible and what at
first seemed low rises progressively higher.*

Eratosthenes, in the 2nd century BC measured the size of the Earth. He
learned that at noon on Midsummer's Day at the town of Syene in Egypt,
the Sun shone directly down a deep well, so that its reflection from the
surface of the water at the bottom could be seen from the top of the well.
In Alexandria at noon on that same day, a vertical stick cast a shadow on
the ground. Eratosthenes saw that it made an angle of 7.2°, or about 1/50 of
a circle (see Fig. 1.1). Assuming that the Earth is a perfect sphere, he used
the angle of the shadow and the distance from Alexandria to Syene, 5,000
stadia, to calculate the earth's circumference as 250,000 stadia. Modern
scholars disagree on the exact length of the *stadium* (*singular of stadia*)
used by Eratosthenes. Values between 500 and 600 feet are suggested. That
puts Eratosthenes' calculation between 24,000 miles and 29,000 miles, in
accord with the modern value, 24,900 miles.

Poisodonius, who lived in the same period as Eratosthenes, determined
the radius by a different method. He took the distance between Rhodes
and Alexandria to be (in modern units) 800 kilometers. In sailing from
one city to the other by the shortest route, the elevation of a certain star
changed by one forty-eighth of a circle. (Although we have no record of
his timing, we must assume that the measurements were made at the same
time of night; for example, a certain number of hours after sunset). Thus
the circumference of the Earth would be 48 times the distance between the
two cities, or 38,000 km, and the radius (equal to $1/2\pi$ of the circumference)
would be $\approx$ 6100 km. Poisodonius' estimate is just $\approx$ 5% low; the correct
value for the mean radius is 6380 km.

---

[a]The Earth's shape is a slightly flattened, "oblate" sphere. Due to the rotation, its
diameter at the Equator is slightly greater than at the Poles.

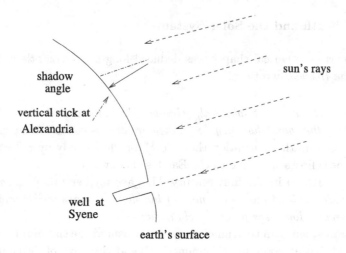

Fig. 1.1   Eratosthenes' measurement of the size of the earth.

The distance to the Sun was first determined by an ingenious geometrical construction. Aristarchus started by assuming that the Moon shines only by reflected light from the Sun (which we know to be true). When the Moon is "half full" the Moon, the Earth and the Sun form a right triangle, as in Fig. 1.2. We can see that the ratio of the short side $R_{\text{Earth}-\text{Moon}}$ to the hypotenuse $R_{\text{Earth}-\text{Sun}}$ is equal to the sine of the angle β. The angle α is directly measurable, and since the total of the interior angles of a triangle is 180°, the angle $\beta = 180° - \alpha$. Therefore, we have all we need to calculate: $R_{\text{ES}} = R_{\text{EM}} / \sin \beta$.

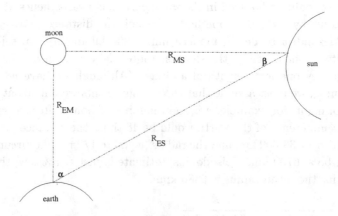

Fig. 1.2   Earth, moon, and sun at half-full moon.

1.3.1 *Parallax*

Parallax means an object's apparent shift of position, when you look at it from different locations, due to the change of viewing angle. If you compare the position of a near object as seen with one eye and then the other, you see that it shifts location against the background. Optical rangefinders use the slight difference of an image's viewing angle from two mirrors a few centimeters apart to measure the distance to an object. One mirror is rotated so that the two images are superimposed, and the rotation angle is calibrated in terms of the object distance.

An early estimate of the Moon's distance was obtained by parallax. The modern value, 365,000 km between the centers of Earth and Moon, is based on the time for a laser pulse to be reflected from the Moon's surface (from a "corner mirror" placed there by the Apollo mission).[b] The Moon's distance changes slightly, due to tides and the Sun's attraction; any one laser measurement has a precision of $\approx$ 3 cm. (See Problem 8.)

Illustrative Example: A rangefinder's two mirrors are $d = 10$ cm apart. In viewing a distant object, the angle of tilt between the two mirrors is $\theta = 3°$.

Solution: To solve, we sketch an isosceles triangle (two equal sides), with the base equal to the distance between the mirrors, and each side equal to the distance between a mirror and the object (see Fig. 1.3). The distance to the object is calculated:

$$L = d/[2\tan(\theta/2)] = 191 \text{ cm}.$$

But we must reduce the number of significant figures because of the limited precision; therefore our result is $L \approx 200$ cm.[c]

---

[b]A "corner mirror" is a set of three plane mirrors set at right angles. Such a mirror will reflect light directly back along its incoming direction, whatever direction that may be.

[c]Distances to stars and galaxies have been measured by using the diameter of the Earth's orbit as the baseline. In 2007 a team of radio astronomers set a record, in the largest distance ever measured by parallax. They measured the distance to the Great Orion Nebula, as 1,350 light years (ly), to an accuracy of 2 percent. The team used a set of radio telescopes named the Very Long Baseline Array (VLBA). The VLBA is a system of 10 radio telescopes, in locations spanning 5000 miles. It provides extremely sharp focus for astronomical objects (the sharpness of focus is equivalent to being able to read a newspaper in Los Angeles while standing in New York), to measure the annual parallax shift (using the diameter of the Earth's orbit as the reference distance). The measurements were carried out over three years, and are accurate to 130 microseconds of arc (3.8×10^{-8} deg.) with respect to a radio-bright galaxy in the background. Reported in *Sky and Telescope* magazine, March 2008.

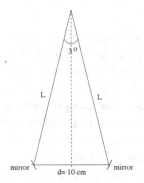

Fig. 1.3

1.3.2   *Galaxies*

A galaxy is a great cluster of stars. Many galaxies, as ours, have the shape of a disk with spiral arms. We see our own "Milky Way" galaxy from the inside, as the wispy white band that is visible overhead on a clear night. It is the fusion of light from many stars, looking through the galaxy toward the edge. The galactic diameter is about 100,000 light years (a light year is the distance traveled by light in 1 year), and it contains about 400 billion stars. We are at the outer edge of one of the arms.

1.4   The Expanding Universe, and the Big Bang

The Universe is expanding; all of the intergalactic distances are increasing. The evidence comes from the *Doppler shift* of light; the same phenomenon that, in sound waves, causes a lower pitch when the source is receding. In 1929 Edwin Hubble assessed his and earlier measurements of the precise wavelengths of light[d] from many galaxies, and discovered that the light from distant galaxies is *red-shifted* (that is, it is shifted to lower frequencies). The more distant the galaxy, the greater is its red shift, which is ascribed to the greater speed of recession. This regularity is known as Hubble's Law; the most recent and precise value for the *Hubble Constant* is 21.5 ± 0.4 km/sec per million light years.

---

[d]The spectra of light emitted or absorbed by gases consist of distinct wavelengths, characteristic of the types of atoms and molecules of the gas (see Ch. 20). If the source is moving toward or away from us the set of wavelength will be "Doppler-shifted" toward shorter or longer wavelengths, respectively.

Illustrative Example: Calculate the recession speed of a galaxy that is 50 light years distant from us.

Answer: The answer is the product of distance times Hubble's Constant (speed/distance), therefore 50×21.5 km/(sec $-$ ly) $= 1,075$ km/s, rounded to 1100 km/s.

If Hubble's Law is traced back in time (this is like running a movie backwards) it seems that all of the galaxies could have begun at a single point and at a definite time in the past, 13.7 billion years ago. This is the prime evidence that inspired the Big Bang theory of the origin of the Universe. The Big Bang theory has been tested in a number of ways. Astrophysicists look at the Universe today, and compare it to calculations of how it would have evolved from a Big Bang, according to the known physical laws. As the great cloud of hot gas and radiation expanded, gravity caused pockets of gas to collapse inward, to form stars and galaxies.

One of the most convincing tests of the Big Bang theory is the prediction that the initially hot ball of radiation and matter would have cooled as it expanded, and 13.7 billion years later, i.e. *today*, all of space would be flooded with the cooled radiation, and this *Cosmic Background Radiation* (CMB) would have a characteristic temperature of a few degrees Kelvin. It was discovered by Arno Penzias and Robert W. Wilson in 1965, who received the Nobel Prize for it. In 2006, instruments on an artificial satellite confirmed the existence of the CMB, and measured its radiation temperature to be 2.725 ± 0.002 K. You can actually "hear" this remnant of the Big Bang as a hiss, if you tune your FM radio to a frequency between stations, or to the frequency of a station that is temporarily not broadcasting.

1.4.1 *Search for extraterrestrial life*

A great discovery in astronomy at the beginning of the third millennium is that there are planets around other stars — lots of planets around lots of stars. But is there life? And if so, is it any sort of life that we can relate to? Well, it can't be a planet of HD 189733b in the constellation Vulpecula, which is 63 light years distant. It's a "hot-Jupiter" planet quite close to a star a bit smaller than our sun; the planet's too hot and too massive for us. Nevertheless, the detection of methane and water in its atmosphere is an important step in the search. Not that the presence of these gases tells us anything about life, rather it demonstrates the technique. As the planet passes in front of its star, the absorption spectrum of the light at the rim of the planet gives its atmospheric composition. The measurement stretched the capability of the Hubble satellite's near infrared spectrometer to the limit, but it should be much easier with the more powerful James Webb Space Telescope, scheduled for launch in 2013.

1.5   A Rough Table of Sizes

An atom's diameter is ≈ 0.1 nm to 1 nm. The thickness of a human hair is $\approx 10^{-2}$ cm. A penny has a diameter of ≈ 2 cm, and a thickness ≈ 0.015 cm. The height of a writing desk is $\approx 3/4$ meter. A kilometer is $\approx 5/8$ mile.

Some astronomical distances:

Earth radius (mean value) 6380 km

Earth to Moon (mean) $\approx 3.85 \times 10^5$ km

Earth to Sun (mean) $\approx 1.50 \times 10^8$ km = one astronomical unit (AU)

Diameter of Solar System $\approx 1.5 \times 10^{12}$ km ≈ 100 AU ≈ 14 light hours

Solar System to Nearest Star $\approx 4.2 \times 10^{13}$ km ≈ 4.5 light years

---

A table of magnitudes and conversion factors

1 Ångstrom (Å) $= 10^{-10}$ m

1 nanometer (nm) $= 10^{-9}$ m

1 meter (m) $= 39.4$ inches $= 3.28$ feet

1 kilometer (km) $= 0.621$ mile

1 astronomical unit (AU) = mean Earth–Sun distance $\approx 1.5 \times 10^8$ km

Speed of light in vacuum $c = 3 \times 10^8$ m/s

1 light year (ly) $= 9.47 \times 10^{12}$ km

Questions

Q1.1. What are the most useful units ($\mathring{A}$, μm, cm, m, km, AU, ly) to express the following sizes:
(a) atoms
(b) people
(c) planets
(d) galaxies

Q1.2. Some artists have pictured a star "cradled" between the horns of a crescent Moon. Why is this impossible?

Q1.3. Most galaxies are receding from us. Does that indicate that we are at the center of the Universe's expansion? Give a reason for your answer.

Q1.4. We have sent a number of space vehicles to explore the nature of other planets. One of their missions is to discover whether they are habitable. What do you think are the most important properties that a planet must have to be our home?

Problems

Please show your work clearly; it will help your reasoning, and, if your answer turns out to be incorrect, will enable us to see how your calculation went wrong.

P1.1. Express your height in:
(a) feet and inches
(b) meters and centimeters
(c) micrometers (μm)
(d) $\mathring{A}$

P1.2. A parallax exercise. Hold a pencil upright in an outstretched hand, and see it change position relative to the far wall of the room when you look with one eye and then the other, while holding your head still. Sketch the arrangement, estimate the angular separation of the two views, and calculate the distance to the wall.

P1.3. Challenge. The Moon's distance has been measured by parallax, by comparing the viewing angles at different times, as the Earth turned. The angle α, known as the *Moon's parallax*, measured at the Equator, from the time that the Moon is rising until it is setting, is about one degree. Make a sketch of the arrangement for making the parallax measurement, and calculate the Moon's distance from these data and the Earth's radius, 6.38×10^6m.

P1.4. Challenge. Estimate the diameter of the Sun, using the known Earth-Sun distance, the diameter and the rate of rotation of the Earth, and your measurement of the time it takes for the Sun to set, from the moment its edge touches the horizon to its complete disappearance.

P1.5. In this problem you can estimate a molecule's size as Pockels did. Assume that the diameter of the oil drop was 3.0 mm, and the area of the film at the Pockels Point was 15 m^2. To simplify your calculation, assume that the droplet and the molecules are shaped as cubes (!), and that they are perfectly packed in the drop (in 3 dimensions) and in the film (in 2 dimensions).

P1.6. The mass of one H$_2$O (water) molecule is $\approx 2.9 \times 10^{-23}$ grams. Assuming that you are composed of pure water, how many molecules are in your body?

P1.7. The distance between San Francisco and Tokyo is 4530 miles. In sailing a great circle route from one city to the other, the elevation of a certain star changes by 65 degrees (at the same time of night). From these data, estimate the circumference of the earth as Poseidonius did. Hint: what fraction of a circle is 65 degrees?

P1.8. A modern measurement of the Moon's distance was made by timing the reflection of a laser pulse from a mirror left on the Moon by one of the space missions. Calculate this time, based on the distance listed in the Table (Sec. 1.5), and the velocity of light.

P1.9. Using Hubble's Law,
(a) calculate the rate of recession of a galaxy that is 2.0 billion light years from us, and compare this speed to the velocity of light.
(b) How much further would a galaxy have to be for its light to fail to reach us? (In other words, a distance, such that the rate of recession would equal the speed of light). This is the size of the known universe.

Answers to odd-numbered questions

Q1. (a) Å; (b) m; (c) AU; (d) ly

Q3. No. In a general expansion, the surroundings of every point are receding.

Answers to odd-numbered problems

P1. Use the conversion factors listed at the end of the chapter.

P3. Sketch should show tangents of Earth's circumference converging at a point on the Moon.
The Moon's distance $d = R/\tan(\alpha/2)$, where R is earth's radius; $d = 7.3 \times 10^8$ m.

P5. Ratio of volume to area $V/A = Nd^3/Nd^2 = d$. In this problem, make sure to put dimensions into the same units (meters). Volume of drop $V = (3 \times 10^{-3} \text{ m})^3$; area of film $A = 15$ m^2; Diameter is ratio $d = 1.8 \times 10^{-9}$ m $= 18$ Å.

P7. 65 degrees is $65/360 = 0.18$ of a circle. Therefore, Poseidonius' method yields the value $4530/0.18 = 2.52 \times 10^4$ miles for the Earth's circumference.

P9. (a) Hubble constant is 21.5 km/s per million light years; therefore, for a distance 2×10^9 ly, the recession speed would be $(2 \times 10^3) \times 21.5 = 4.3 \times 10^7$ km/s.
(b) For a recession speed equal to the speed of light, $c = 3 \times 10^5$ km/s, the distance would be a factor 150 times greater, i.e. 300×10^9 ly.

Chapter 2

Linear Motion

2.1 Introduction

In ancient Greece the physics of Aristotle dealt with change, more for its causes and effects than for a description of movement itself. Motion was compounded out of a succession of positions, rather than as a smooth rate of change, a speed. These ideas are tested by Zeno's Paradox, which supposes a race between Achilles and a tortoise. The paradox "proves" that if the tortoise were given a head start, *no matter how small*, Achilles could never catch up. Here is how the paradox is presented.

> *Achilles was a superb athlete, so to make the race more competitive, the tortoise was given an advantage, by starting at the halfway point. Achilles and the tortoise began at the same time. In the first "stage", Achilles ran to the mark where the tortoise had started, but the tortoise had advanced to a new position; then in the second stage Achilles reached that position, but the tortoise had again moved on. And so it continued; at each stage Achilles reached the place where the tortoise had been, but the tortoise was no longer there. Thus, it was argued, Achilles could never catch up to the tortoise.*

The paradox is due to the assumption that the time intervals for the infinite series of displacement "stages" would sum up to an infinitely long time. But actually, the time intervals for each stage get shorter and shorter, and although they form an infinite series, the series sum is finite.[a] We can

---

[a]Let the speeds of Achilles and the tortoise be V and v, respectively; the course length is L and the tortoise begins at $x = L/2$. The time for Achilles to reach the halfway point is $L/2V$; in that time the tortoise has advanced a distance $vL/2V$. The time for Achilles to reach this point is $vL/2V^2$, etc. Proceeding in this way, we obtain the following series

avoid the paradox completely, by viewing the motion not as a series of stages but as smooth and continuous movement, and just compare the speeds. Since we are accustomed to travel on bikes, in cars and other vehicles, we have no difficulty in understanding continuous motion.

2.2 The Kinematics of One-Dimensional Motion

The science of motion is *kinematics*, involving only two quantities: position and time. Suppose that an object moves an extended distance, from x_1 at time t_1 to x_2 at time t_2. Its average speed is

$$\bar{v} = \frac{x_2 - x_1}{t_2 - t_1} = \frac{\Delta x}{\Delta t} \qquad (2.1)$$

By contrast, the *instantaneous* velocity v is the same ratio, $\Delta x / \Delta t$ for a very, very short interval of time,

$$v = \lim_{\Delta t \to 0} \Delta x / \Delta t . \qquad (2.2)$$

We have illustrated the difference between $\bar{v}$ and v in Fig. 2.2. The instantaneous velocity is the local *slope* of the path at the point C. The slope is the tangent to the curve. In Fig. 2.1, it is $(x_2' - x_1')/(t_2' - t_1')$.

Fig. 2.1   Motion as seen from a train — courtesy Wikipedia.

2.3 Proportions

Many physical quantities are *proportional to each other*, i.e. increase and decrease at the same rate. When A is doubled, so is B; when A is increased

---

for the total time: $t = (L/2V)[1 + v/V + (v/V)^2 + \cdots]$ The series has a finite sum: $t = (L/2V)/(1 - v/V)$, which we give without proof (it requires calculus). From the series sum you may see that, as Achilles' speed becomes more and more dominant, his time to catch up to the tortoise tends to the value $L/2V$.

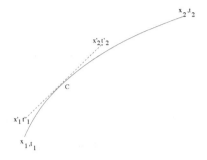

Fig. 2.2   Average and instantaneous velocities.

by 10 percent, so is B. An equivalent phrase is *A varies linearly with* B. Thus, when speed is constant, distance is *proportional to* the time of a trip.

2.4   Acceleration

An object's speed may change during an interval; at any moment the speed may be greater or smaller than the average. So we distinguish *instantaneous speed* v from the *average speed* $\bar{v}$, as illustrated in Fig. 2.2.

The rate of change of v *is the acceleration* a; if the instantaneous speed is initially v_1 at t_1 and it changes to v_2 at t_2,

$$a = \frac{(v_2 - v_1)}{(t_2 - t_1)}, \; or \; a = \frac{\Delta v}{\Delta t}. \tag{2.3}$$

If the initial velocity is v_0 at $t = 0$, then

$$v - v_0 = at. \tag{2.4}$$

Thus we have that velocity v at time t is $v = v_0 + at$ and $\bar{v} = \frac{v_0 + (v_0 + at)}{2} = v_0 + \frac{1}{2}at$.

Now we can obtain a formula relating the position to the elapsed time

$$x = x_0 + \bar{v}t = x_0 + v_0 t + \frac{1}{2}at^2. \tag{2.5}$$

The factor 1/2 arrives because the average over the interval 0 to t is $\bar{v} = \frac{1}{2}(v_o + v)$. Figure 2.3 illustrates how position, speed and acceleration are graphically related. The speed is the slope of position versus (vs.) time, and the acceleration is the slope of the velocity vs. time.

Acceleration can be positive or negative; if positive, the speed is increasing, and if negative it is decreasing (or increasing in the negative direction). When the magnitude of velocity is decreasing, we say that the object is *decelerating*.

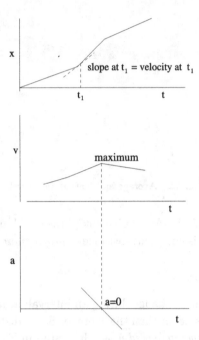

Fig. 2.3 Plots of position, velocity, and acceleration over an extended time interval. Note that v is the slope of x vs. t, and a is the slope of v vs. t.

It is interesting to compare the order of magnitude of the speeds and accelerations of various objects. A fast pitch baseball travels at about 90 miles/hour; in metric units, 40 m/s. The Earth, in its orbit about the Sun, has an average speed nearly a thousand times larger, 38 km/s. Slightly faster is the average speed of an oxygen molecule at room temperature; $\approx$ 50 km/s. And the recession velocity of a galaxy near ours is greater still, $\approx$ 200 km/s. A car's acceleration after completing a stop at a traffic light, might be $\approx$ 4 m/s^2. A drag racer's acceleration, is much greater (see Challenge Problem). A bullet, as it travels through a rifle barrel has $a \approx 10^6$ m/s^2.

2.5   Kinematics of Two- and Three-Dimensional Motion. Vectors

For motion in a plane or in three-dimensional space it is convenient to use the Cartesian coordinate system of mutually perpendicular x, y, and z axes. To combine all three directions in position, velocity and acceleration, we can use *vector notation*. Vectors are useful in describing other quantities such as forces, and we will apply vectors in several forthcoming chapters. A vector quantity has both magnitude and direction; it is the same vector, no matter where it is located. In contrast, a *scalar* has magnitude only. Graphically, a vector is shown as an arrow pointing in the proper direction, with a length indicating its magnitude. In writing or typing, a vector is denoted in boldface (e.g., **A** or with an arrow over it, e.g., $\vec{A}$. Vectors can be added graphically by translating them (keeping magnitudes and directions unchanged), to connect, head-to tail. The sum of two vectors **B** + **C** = **D** is the vector connecting the tail of one to the head of the other, as in Fig. 2.4. The negative of a vector has the same length, but with its direction reversed.

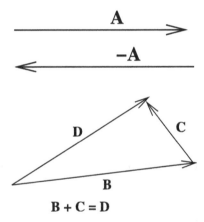

Fig. 2.4   Illustration of the vector **A** and its negative, and the addition of vectors **B**, and **C**.

The addition of two perpendicular vectors produces a magnitude given by the Pythagorean theorem: if **A** and **B** are perpendicular, $\mathbf{C^2} = \mathbf{A^2} + \mathbf{B^2}$. Vectors can also be added (and subtracted) by adding (and subtracting) their components. This can be done algebraically, rather than graphically. For example, if **A** + **B** = **C**, then the components of **C** are $C_x = A_x + B_x$; $C_y = A_y + B_y$; $C_z = A_z + B_z$. The components of a two dimensional vector are shown in Fig. 2.5. In 3D rectilinear (straight line) motion, the kinematic equations are the vector combinations of Equations (2.2), (2.3) and (2.4) in each of the three dimensions.

$$a_x = \frac{(v_x - v_{ox})}{t} \;;\quad a_y = \frac{(v_y - v_{oy})}{t} \;;\quad a_z = \frac{(v_z - v_{oz})}{t} \qquad (2.6)$$

$$x = x_o + v_{ox}t + \frac{1}{2}a_x t^2 \;;\quad y = y_o + v_{oy}t + \frac{1}{2}a_y t^2 \;;\quad z = z_o + v_{oz}t + \frac{1}{2}a_z t^2 \qquad (2.7)$$

Now the three 1D equations can be represented by one compact vector equation:

$$\mathbf{r} = \mathbf{r_o} + \mathbf{v_o}t + \frac{1}{2}\mathbf{a}t^2 \,, \qquad (2.8)$$

where $\mathbf{r} = x\hat{x} + y\hat{y} + z\hat{z}$, and similarly for $\mathbf{v}$ and $\mathbf{a}$. The symbol $\hat{x}$ is a vector of unit length in the x direction; similarly for $\hat{y}$ and $\hat{z}$. Figure 2.5 illustrates how a two dimensional vector is analyzed into its components.

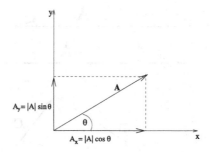

Fig. 2.5   A two dimensional vector and its rectlinear (x- and y-) components.

2.6   Relative Motion

Suppose that you are in a boat on a river with a strong current. Your boat speed is relative to the water, but the water moves relative to the shore. If you are trying to go upstream and the speed of your boat is less than that of the water, then you will move backwards relative to the shore.

We need to describe motion relative to a frame of reference, e.g., the water or the shore. If the boat moves downstream at a speed v_w relative to the water, and the water's speed is u relative to the shore, then the boat's speed, v_s is larger relative to the shore than its speed relative to the water, $v_s = v_w + u$. If the boat were going upstream, its speed relative to the shore would be $v_s = v_w - u$. In both cases we have taken the speed of the boat relative to the water as positive.

We have used one-dimensional motion for illustration purposes. If the boat wants to go from one shore to the other, then it will be carried downstream and has to head upstream if it wants to reach the opposite shore directly opposite its starting point. With vectors we can write this succinctly as

$$\vec{v}_s = \vec{v}_w + \vec{u}. \tag{2.9}$$

This equation also applies when the boat travels up or downstream.

Illustrative Example: A boat that travels 20 m/s relative to the water heads directly across a stream moving at 15 m/s. The opposing shore is 40 m away.

(a) How far downstream will the boat be when it reaches the opposite shore?

(b) How far will the boat have traveled?

(c) What is the speed of the boat relative to the shore?

(d) How long does the crossing take?

Answer:

(a) $\tan \theta = 15/20 = 0.75; \theta = 36.8°; y = 40 \tan \theta = 30$ m

(b) The boat will have gone $\sqrt{(40)^2 + (30)^2} = 50$ m

(c) The speed of the boat relative to the shore is $\sqrt{(20)^2 + (15)^2} = 25$ m/s

(d) The time of travel is 40 m/(20 m/s) = 2 s or 50 m/25 m/s = 2 s

Questions

Q2.1. Suppose that $v = 0$ at $t = 0$; can acceleration have a finite value at $t = 0$?

Q2.2. Can the acceleration $\vec{a}$ be non-zero if $a_x = 0$? Is the reverse possible, i.e., can a_x be zero if $\vec{a} = 0$? Explain.

Q2.3. Can $\vec{v} = 0$ if $\vec{a}$ is not 0? Is the reverse possible, that is, can $\vec{a}$ be zero if $\vec{v}$ is not 0? Explain.

Q2.4. Can the velocity $\vec{v}$ be zero if $\vec{a}$ does not equal 0? Is the reverse possible? Explain.

Q2.5. Can $\vec{v} = 0$ if $\vec{v}$ is not zero? Explain.

Q2.6. Can you add $\vec{v}$ and $\vec{a}$? Explain.

Q2.7. Give at least two examples of scalars and two examples of vectors.

Q2.8. Figure 2.6 shows the velocities of persons A and B. They cross at a time t_1. Which person will have gone further, if either, by the time t_1? Give a reason for your answer.

Q2.9. Can a boat's speed relative to the shore be negative? Explain.

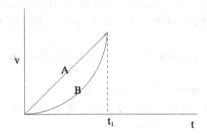

Fig. 2.6   The speed of two persons from $t = 0$ to t_1.

Problems

Be sure to show all your work.
Note: A convenient conversion: 1 mile/hour (mph) = 0.447 m/s.

P2.1. Estimate your acceleration as you take the first step to begin walking at your normal pace. Estimate what speed you would have attained if you maintained that acceleration for 1 minute. Why can't you do that?

P2.2. A train's trip, along a straight track, has the following history. It starts with an acceleration of 0.10 m/s^2 for 3.0 minutes, then zero acceleration for 30 minutes, and finally decelerates at a rate of -0.20 m/s^2 until it comes to a complete stop. (a) What is the train's speed after 1 minute? (b) What is the train's maximum speed? (c) How far did it travel over the entire trip?

P2.3. Two trains begin traveling 10 miles apart on the same track, each at a speed of 20 miles/hour. Due to a signalman's error, they are heading toward each other, on the same track. (a) What is the total time, from start to crash? (Ignore the time to accelerate). (b) A bird begins flying at a speed of 50 mph at the starting time the trains start, shuttling from one train to the other, until it is crushed when the trains crash. What was the bird's total flight distance?

P2.4. A boat's speed upstream on a river is 4.0 miles/hour, and downstream is 7.0 miles/hour (measured relative to the shore). In a certain trip it travels from its base to a port 12 miles upstream, and immediately turns around and travels downstream to its base.
(a) How long does it take to make the round trip?
(b) What is the boats average speed for the whole trip?
(c) What is the speed of the river, relative to the shore?

P2.5. As car A starts, it is passed by another car (B) traveling at a velocity of 12 m/s in the same direction. Car A accelerates at 3.5 m/s^2 from rest. At what time does car A pass car B?

P2.6. A candy vendor walks from car to car through a train, from one end to the other; as soon as the vendor gets to one end, she turns and walks toward the other. The vendor walks steadily at 3 km/hour in the train's frame of reference; the train is 100.0 m long and it is traveling at 50 km/hr. How far does the vendor walk relative to the ground during a train's trip of 100.0 km?

P2.7. In his race with the tortoise over a 1.0 km course, Achilles gives the tortoise a head start of 0.5 km. Each travels at constant speed, Achilles at 20.0 km/hr and the tortoise at 20.0 m/hr. (a) How long does it take Achilles to reach the tortoise? (b) How far had he traveled?

P2.8. As you step into an elevator on the first floor you press the button for the 4th floor and the elevator immediately starts accelerating, reaching its maximum speed, 1 m/s, in 3.0 seconds. When it approaches the 4th floor it decelerates at a rate equal and opposite to its acceleration at the start. The distance from the first to the fourth floor is 50.0 meters. How much time for the trip?

P2.9. A dog wanders from home 561 m due E and then 348 m at 63° N of W. Find the direction and distance the dog must go to travel home by the shortest route. Do so graphically and algebraically. You can give the answer in terms of the distance in the E-W and S-N directions.

P2.10. Two trains are heading toward each other on the same track. One is traveling at 50 miles/hour, and the other at 60 miles/hour. How much time elapses from the moment they are 10 miles apart, until they collide?

P2.11 A car is traveling at a steady speed of 60 mph; it slows down to 42 mph in 6.0 minutes. What is the car's acceleration and how far did it travel during the slow-down process?

P2.12. Estimate your average and maximum instantaneous speeds while commuting to school.

P2.13. A car starts from rest with a constant acceleration. In an interval of time Δt the distance traveled increases by a factor of 3. During that interval of time how much does the speed increase?

P2.14. Estimate the rate your fingernails grow, *in km/hour*.

P2.15. Vectors **A** and **B** are drawn on the blackboard. Vector **A** has a magnitude of 5 units, and points vertically up; Vector **B** has a magnitude of 10 units and points horizontally to the right (to 3 o'clock on a clock face). What are the magnitude and direction of $\mathbf{A} + \mathbf{B}$?

P2.16. Challenge. A drag race is a competition between cars: (beginning from $v_o = 0$ at $t = 0$), to reach the quarter-mile (1320 feet = 0.402 km) finish line in the shortest time. The winner in a recent contest had an elapsed time 4.462 s. Calculate the winner's final speed, assuming constant acceleration. See the Note before Problem 1, for a convenient conversion factor.

P2.17. A passenger walks at 55 mph (mph = miles/hour) relative to the ground on a train traveling at 60 mph. What is the passenger's velocity relative to the train?

P2.18. A boat driver wants to reach the opposing shore 20.0 m away directly opposite her starting point. The boat's speed is 12 m/s and the water's speed is 10.0 m/s. (a) What is the direction that the boat must head in?
(b) How long will the trip to opposite shore take?

Answers to odd-numbered questions

Q1. Yes. In fact, the only way for motion to begin is with a finite acceleration.

Q3. Yes, the velocity can go through zero; yes for a constant velocity, $\vec{a} = 0$.

Q7. Time and speed are scalars; velocity and acceleration are vectors.

Q9. No, the speed is never negative. The velocity can be negative.

Answers to odd-numbered problems

P3. (a) One-quarter hour. (b) 50 mph × 1/4 hour = 12.5 miles.

P5. The distance covered by car A is $d = \frac{1}{2}at^2 = v_B t$ when car A passes car B. Thus $t = 2v_B/a = (24 \text{ m/s})/(3.5 \text{ m/s}^2) = 6.9$ s.

P7. Let t = time (in hours) for Achilles to catch the tortoise. In that time, the tortoise travels $20t$ meters, and Achilles travels $(500 + 20t)$ m, in $(500 + 20t)/(2 \times 10^4)$ hours. Equating, $t = (500 + 20t)/(2 \times 10^4)$. Solving for t, $t = 500/(2 \times 10^4 - 20) = 0.025$ hr = 1.5 mins.

(b) He traveled 500.5 m in $t' = 1.5$ minutes. Thus, it is as if the tortoise was not moving.

P9. Along E-W: $561 - 348\cos(63)° = 403$ m; along N-S: $348\sin(63)° = 310$ m.

P11. Acceleration = -18 mph/$(1/10$ hr$) = 180$ mph/hr. Distance covered is $v_0 t + 1/2at^2 = 60(1/10) - (1/2)(180 \text{ mph/hr})(1/100 \text{ hr}^2) = 5.1$ miles.

P13. $v = at$, and $d = (1/2)at^2$. Therefore v increases by $\sqrt{3}$.

P15. The magnitude of $\mathbf{A} + \mathbf{B} = \sqrt{A^2 + B^2} = \sqrt{125} = 11.2$, the sum vector points upward and to the right, at an angle $\tan^{-1}(0.5) = 26.6°$ above horizontal.

P17. The passenger must walk opposite to the train's direction at $60 - 55$ mph = 5 mph, or at a velocity $55 - 60 = -5$ mph, if the train's direction is taken as positive.

Chapter 3

Galileo; Free Fall, Projectile Motion

3.1 Galileo's Observations vs. Aristotelian Ideas on Motion

After our study of kinematics we face the causes of motion of physical objects; *what makes them move, how fast do they go, and how do different bodies react?*

Aristotle, who was concerned with purpose ... *teleology* ... argued that all bodies had natural locations, and they tended toward their predestined order. If an object was displaced from its proper location, it "wanted" to return, and there to remain. Heavy bodies had a "natural" motion toward the center of the Earth. All other motion was "violent", because it required a constant driving force. Except for falling, things moved only when pushed, and when pushing stopped, so did the object. However, circular motion could be eternal ... indeed, "divine". Dante called Aristotle *The Master of Those Who Know.* If one wished to know, the way to knowledge was by a careful reading of Aristotle's texts; one must study the meaning of difficult passages, debate them, and read the extensive dissertations that had been written on their implications.

In the 17th century a series of books appeared which began to demolish the rule of Aristotelian philosophy. The authors were Francis Bacon, Rene Descartes, William Gilbert and Galileo. Their crucial departure from earlier philosophy was their focus on observation and experimentation. Instead of inventing reasons *why* an effect existed, they studied *how* it depended on time, distance, and other physical quantities.

Galileo was born in 1564 in Pisa, the oldest of seven children. His father was a music teacher, an unusual man, in that he had an interest in mathematics. For example, he tried to invent a musical scale related to the heavens, the "music of the spheres". As we shall see, this interest, more than his father's explicit desires, may have influenced Galileo's choice of career. Galileo entered the University of Pisa when he was seventeen, intending to follow his father's wish that he study medicine. In his early years he earned a reputation for contradicting his professors. He sat in on some lectures on geometry, which so intrigued him that he defied his father, and switched course to study mathematics. For a few years after leaving the university he tutored students in mathematics. About 1587 he began applying mathematics to a series of physical problems. He proposed a practical approach to locating the centers of gravity of certain solids, an advance which earned him his first recognition abroad. In 1589 he was appointed to the Chair of Mathematics at the University of Pisa. Three years later, on the strength of his teaching in Pisa and with the support of patrons, he was appointed Professor of Mathematics at the University of Padua. For a year he was mainly occupied with practical problems, such as artillery ranging. Then he began a study of accelerated motion and falling bodies, experimental observations and measurements that became the foundations for classical mechanics. He devised a way to study *diluted gravity* by experiments on an inclined plane. A ball was started from rest on a very gently sloping plane (less than 2°), and its positions were marked at regularly periodic times. His method of timing would have delighted his father: he judged the intervals by musical beats. Galileo discovered that the speed increased proportional to the time, and the distances increased proportional to the speed; thus, the *acceleration, the rate of change of speed, was constant, and the distances increased as the square of the time.*

He had been convinced since his youth that all dense bodies, of whatever weight, fall to earth with the same acceleration. In his later years he wrote that during a hailstorm, he had seen that hailstones of different sizes fell to the ground at the same time, and he reasoned that they must have formed and begun falling from the same height.[a]

His new studies confirmed his belief that the *acceleration due to gravity is the same for all bodies.* He knew that this is not strictly correct, especially

---

[a]It is an interesting analysis, but we disagree. It is likely that most of the hailstones had reached their terminal velocities (the limiting speeds that are caused by air resistance; see Ch. 4, Sec. 4.6) before they hit the ground. Therefore, during at least part of their fall, the hailstones were falling at speeds that depended on their size, so that those that hit the ground simultaneously must have started at different times and heights. We have to conclude that Galileo's inspiration for his famous discovery, that, barring air resistance, all bodies fall with the same acceleration, which is true, was based on a false premise.

Galileo Galilei

Portrait of Galileo Galilei by Giusto
Sustermans

Fig. 3.1   Portrait of Galileo, courtesy Wikipedia.

for very light bodies, but he reasoned that it would be true in the absence
of air resistance.

Actually, the acceleration of gravity varies slightly with altitude and
latitude, but for most purposes one can ignore the variations. It is convenient to use an average value, the *standard acceleration of gravity g*,
$g = 9.80$ m/s^2 ≈ 32 feet/s^2. There is a unit of gravitational acceleration:
1 Gal (after Galileo) $= 0.01$ m/s^2, in terms of which, the standard value
for g is 980 Gals. Sensitive modern instruments can measure variations as
small as 1 μGal $= 1 \times 10^{-6}$ Gal. These instruments are used for mapping
variations due to differences in latitude, altitude, or local differences in the
composition of the Earth.

If you throw an object (e.g., a ball) upwards, it will slow down due to
gravity and come to a momentary stop at the height h before falling back
to earth. If $y > 0$ upwards,

$$v = v_0 - gt$$

$$y = v_0 t - \frac{1}{2}gt^2 \tag{3.1}$$

At the top $y = h$ and $v = 0$, so the time to reach the top is $t = v_0/g$ and $y = h = v_0 t - \frac{1}{2}\frac{v_0^2}{g} = \frac{1}{2}\frac{v_0^2}{g}$.

Illustrative Example: A ball is thrown upward at the ground with a velocity of 15 m/s. (a) What is the time to reach the top? (b) What is the height of the top (h)? (c) What is the time taken for the entire trip? (d) What is the velocity at 2 s? Take $g = 10$ m/s^2.

Answer: (a) The time is $v_0/g = (15$ m/s$)/(10$ m/s$^2) = 1.5$ s; (b) $h = (1/2)v_0^2/g = (1/2)(225$ m^2/s$^2)/(10$ m/s$^2) = 11.25$ m ≈ 11 m; (c) The time for the entire trip is twice that to reach the top or 3 s; (d) At 2 s, the ball is on its way down; since it starts at the top with no speed, its velocity will be $v = -gt$, where t is the time since it left the top, or 0.5 s; $v = -(10$ m/s$^2)(0.5$ s$) = 5$ m/s. Alternatively, $v = v_0 - gt = 15$ m/s $- (10$ m/s$^2)(2$ s$) = -5$ m/s.

3.2   Projectile Motion

In Chapter 2 we studied how the kinematic equations combine for the general case of three dimensional motion. The path of *projectiles* is a specific case of two dimensional motion, that was first analyzed by Galileo. Before his analysis, there were some strange ideas about the motion of projectiles. One of Galileo's most important insights was the realization that moving objects slow down and stop only because of some impediment such as friction. He turned the classical idea on its head: In place of the old notion that a body needed a motivating force to continue moving, he proposed that, without any external forces, a body in motion would continue, in a straight line, forever.

He wrote that a projectile's path . . . its *trajectory* . . . is composed of two simultaneous motions: constant horizontal speed combined with vertical gravitational free fall. His graphical construction showed that the trajectory is a *parabola*. Galileo's geometrical treatment spans several pages, but we can show it more economically by algebra. Begin with the set of horizontal (x) and vertical (y, positive upward) kinematic equations:

$$x = x_o + v_{ox}t \tag{3.2}$$

$$y = y_o + v_{oy}t - 1/2gt^2 \tag{3.3}$$

Since the acceleration is down, it is negative.

Let us simplify the equations, setting $x_o = 0$ and $y_o = 0$. Then we solve Eq. (3.2) for t, and subsitute it in Eq. (3.3):

$$t = x/v_{ox}; y = v_{oy}(x/v_{ox}) - \frac{1}{2}g(x/v_{ox})^2 \qquad (3.4)$$

Rewriting Eq. (3.4) shows a parabolic dependence of y on x:

$$y = ax + bx^2, \qquad (3.5)$$

where $a = (v_{oy}/v_{ox})$, and $b = -g/(2v_{ox}^2)$. Another convenient form of the trajectory is obtained in terms of the total initial velocity $\mathbf{v_o}$ and the 'elevation angle' θ, the angle that $\mathbf{v_o}$ makes with the horizontal. Then $a = \tan\theta$, $b = -g/[2(v_o\cos\theta)^2]$. Figure 3.2 shows a few parabolic trajectories with varying velocities and elevation angles; Fig. 3.3 shows the horizontal and vertical components of projectile motion.

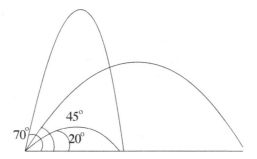

Fig. 3.2   Projectile paths for different launch angles.

Illustrative Example: An arrow is shot at 30° relative to ground at a speed of 12 m/s. (a) Find the height it reaches and (b) the horizontal distance it travels by the time it reaches its highest point.

Answer: (a) $v_{0y} = 12\sin 30° = 6$ m/s; $t = (0 - v_{0y})/g = 0.612$ s; $y = v_{y0}t - (1/2)gt^2 = 5.5$ m;
(b) $x = v_{0x}t = v\cos 30°t = 6.4$ m

3.3   Range of a Projectile

In gunnery, or *bocce*, pitching pennies and horseshoes, one wants to send a projectile a certain distance; a predictable *range* R. Here we transform

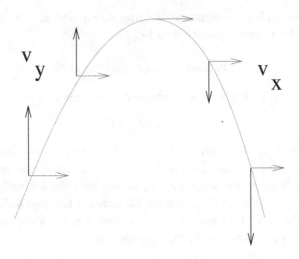

Fig. 3.3   Horizontal and vertical components of the velocity of a projectile.

Eqs. (3.2) and (3.3) to a convenient expression for R, in terms of the projectile's initial velocity $\mathbf{v_o}$ and the *angle of inclination* θ.

First, we recognize that $y = 0$ at the beginning and at the end of the projectile's flight. Then setting $y = 0$ in Eq. (3.5), we get

$$y = 0 = ax + bx^2 .$$

Factoring, the equation can be written

$$0 = x(a + bx) .$$

We see that there are two solutions for x: $x = 0$ and $x = -a/b$. The first corresponds to the projectile's starting point, and the second is the endpoint; the *projectile's range* R. Substituting the values of a and b from the line following Eq. (3.5), we get

$$R = -a/b = 2v_{ox}v_{oy}/g . \tag{3.6}$$

The x and y components of $\mathbf{v_o}$ are:

$$v_{ox} = v_o \cos \theta ; \quad v_{oy} = v_o \sin \theta ,$$

where θ is the angle with respect to the horizontal (ground). Substituting in Eq. (3.6), we get

$$R = \frac{2v_o^2}{g} \sin \theta \cos \theta$$

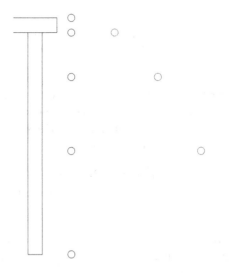

Fig. 3.4 Two balls falling from a table; one ball falls straight down and another falls with a horizontal velocity at the same time. The two balls are shown at constant time intervals.

With the trigonometric relation, $2\sin\theta\cos\theta = \sin(2\theta)$ we get

$$R = \frac{v_o^2}{g}\sin(2\theta)\,. \tag{3.7}$$

Consider how R changes with the inclination angle; R increases as θ grows from 0 to 45°, and then declines to 0 as θ reaches 90°. The maximum range at 45° is the very simple expression

$$R_{\max} = v_o^2/g\,.$$

Of course, these formulas correspond to zero air resistance, but they are reasonably accurate for compact bodies at low speeds, as in some sports such as the hammer throw and the broad jump. ("Low speed" means much less than the velocity of sound).

Illustrative Example: An athlete prepares to make a broad jump by running at a speed of 25 miles/hour, and launching himself at an upward angle of 45 degrees. How many meters is the length of the jump?

Answer: The first task is to convert the speed into meters/sec, and then to apply the range formula Eq. (3.7). The conversion factor (from Ch. 2) is 1 mph = 0.442 m/s; the runner's speed is 11 m/s. Therefore the jump distance is 12.3 m, or to the desired accuracy 12 m.

Questions

Q3.1. How does gravity cause the bubbles in soda to rise, instead of falling down?

Q3.2. You drop your keys while riding in an elevator that is traveling up, but decelerating as it approaches the destination floor. Is the time for the keys to hit the floor (a) longer, (b) shorter, (c) the same, as the time for its fall when the elevator is not moving? Explain.

Q3.3. Compare quantitatively the up and down motions for a ball thrown upwards, e.g. the velocities at a given height, the time for up and down motions, etc.

Q3.4. Ball A is thrown downwards with a speed v_0. Ball B starts from rest at the same time. Compare $v_A - v_B$ at the start and at a later time. Explain.

Q3.5. Is the acceleration of a ball thrown upwards zero at the top of its path? Explain.

Q3.6. Ball A rolls off a table at a horizontal speed v_0. Ball B rolls off the same table at the same time at zero speed.
(a) Which ball reaches the ground first, if either? Explain.
(b) Which ball has a larger speed when it reaches the ground? Explain.

Q3.7. A ball is thrown upward at an angle of 30° to the horizontal. Is the ball's speed = 0 at the top of its trajectory? Explain.

Q3.8. A key falls from your pocket. Will the distance fallen in the first 0.1 s be the same as that covered in the next 0.1 s? Explain.

Problems

P3.1. (a) When you throw a stone up, what is its acceleration immediately after it leaves your hand?
(b) What is its acceleration at the top of its travel?
(c) What is its velocity at the top of its travel?

P3.2. You launch an arrow vertically upward with initial velocity $v_{oy} = 8.0$ m/s.

(a) What is its upward velocity at maximum height?
(b) What is its maximum height?
(c) How much time did it take to reach maximum height?
(d) What is the arrow's maximum downward velocity, and when does it occur?
(e) Compare the arrow's speed halfway up and halfway down.

P3.3. A daredevil motorcyclist plans to "jump" across the Grand Canyon. He chooses a spot where the canyon is only one-half kilometer wide and the two sides are at the same altitude. A ramp is prepared, inclined at 45° to give maximum range. How much speed must he have when he takes off from the ramp, to reach the other side? Assume that the cycle and rider have excellent streamlining, so that air resistance can be neglected (*not really possible in real life*).

P3.4. A marksman aims at the center of a target that is 100.0 m distant; both marksman and target are on platforms 10.0 m high. The bullet travels at a speed of 400.0 m/s.
(a) By how much distance does the bullet miss the center of the target?
(b) If the target began to fall at the instant the gun was fired, where would the bullet hit?

P3.5. Astronauts on the Moon pretended to play golf there. Since Moon's gravity is ≈ 0.16 of Earth's, how much greater is the maximum range of a golf ball on the Moon than on Earth?

P3.6. Neutrons emerging from a research reactor have a velocity $v_x = 1200$ m/s. How far do the neutrons fall while traveling a horizontal distance of 200 m?

P3.7. Superman is supposed to leap over tall buildings. Calculate his minimum initial speed and his acceleration during his spring, for a leap over the Space Needle (184 m high). Estimate his change of height, from a crouch to full extension, as 1.0 meter. Assume zero air resistance.

P3.8. Challenge. This really happened: a salmon broke a building's second story window. An eagle was carrying a salmon that it had just snagged from the river. Another eagle attacked it to get the fish. During the fight the fish fell in an arc and crashed through the window. Here we reconstruct the accident, with some plausible data. The window was 10.0 m above the ground; the eagle was flying at 2.0 m/s, at a height of 50.0 meters, along a course directed at 90° to the plane of the window. (It will help to make two sketches of the course and the window; a horizontal and a vertical view). How far from the window (horizontal distance) was the eagle when it released the salmon?

P3.9. Show that the ranges of a projectile are the same for throws at an angle θ and $(90 - \theta)$.

P3.10. A test of one's reaction time can be carried out with a friend and a meter stick. The friend holds the meter stick hanging straight down from one end. You position your hand midway down the stick, with your thumb and forefinger at the 50.0 cm mark, ready to grab the stick. Your friend releases the stick without warning, and you grab it as soon as you see it begin to fall. Calculate your reaction time from the equations of motion of free fall and how far it fell before you grabbed the stick.

P3.11 A ball falls out of a small plane traveling horizontally at a speed of 100.0 m/s at an altitude of 1000.0 m.
(a) How long does it take for the ball to reach the ground, if you can neglect air resistance?
(b) What is the range of the ball when it hits the ground?
(c) Find the horizontal and vertical velocities of the ball just as it reaches the ground.

P3.12. A ball is thrown upwards at an angle of 60° to the horizontal (from the ground) at a speed of 20.0 m/s.
(a) How high will the ball go?
(b) What is the range of the ball?
(c) What is the time taken for the ball to return to ground?
(d) What other angle of throw (if any) will result in the same range?

Answers to questions

Q1. The bubbles rise because they are lighter than the liquid.

Q3. The motion up and down are symmetrical in all aspects.

Q5. The acceleration does not change; it is $= -g$.

Q7. No, since the horizontal velocity does not change.

Answers to problems

P1. (a) The normal acceleration of gravity, g.

(b) the same answer as in (a).

(c) $v = 0$.

P3. We use the range formula for a $45°$ launch, $R = v^2/g$. With $R = 500$ m, the minimum speed required is $v = \sqrt{Rg} = 70$ m/s $= 252$ km/hr.

P5. A factor $1/0.16 \approx 6.3$ greater.

P7. To relate his initial speed v_i to the height h of his leap, we have the relation $h = v_i^2/2g$. He would need an initial speed of 60 m/s (216 km/hr) to spring 184 m high. To calculate his needed acceleration, we use $a = (v_f^2 - v_i^2)/2y = (60 \text{ m/s})^2/(2 \text{ m}) = 1800 \text{ m/s}^2$.

P9. $R = \frac{v_0^2}{g}\sin(2\theta) = \frac{v_0^2}{g}\sin(180° - 2\theta)$.

P11. (a) $y = -\frac{1}{2}gt^2$. Here $y = -1000$ m. Take $g = -10$ m/s^2; then $t = \sqrt{2y/g} = \sqrt{2000/10} = 14.1$ s.

(b) $x = v_x t = 100$ m/s $\times 14.1$ s $= 1410$ m.

(c) $v_x = 100$ m/s; $v_y = gt = -10$ m/s$^2 \times 14.1$ s $= -141$ m/s.

Chapter 4

Newtonian Mechanics

We might say that in the previous chapters we have described the "how" of motion. This chapter deals with the "why" of motion.

Isaac Newton, the prime creator of classical mechanics, was born in 1642, the same year that Galileo died. He grew up in a small village, where he tended sheep (indifferently), and showed little intellectual brilliance. He entered Cambridge at the age of nineteen, a lonely student with little money. In his biography of *Isaac Newton*, James Gleick stated:

> He had enough for his immediate needs: a chamber pot; a note-
> book of 140 blank pages, three and a half by five and a half
> inches, with leather covers; a quart bottle and ink to fill it; can-
> dles for many long nights, and a lock for his desk.

The curriculum was dull, but he had access to the library, where he read Aristotle, Descartes and Galileo. In his notebook he began to pose questions about matter and motion, light and the cosmos. In 1664 an epidemic of bubonic plague struck Europe and then England. The Cambridge colleges were forced to shut down, to let the students and fellows disperse into the countryside. Newton came home, to study by himself. He obtained a larger notebook and began to record thoughts that evolved into original research. It is often stated that he sat under an apple tree and that the fall of an apple led to his theories of motion. It is not known whether there is any truth to this story.

> He set himself problems; considered them obsessively; calculated
> answers, and asked new questions. He pushed past the frontier
> of knowledge (though he did not know this). The plague year
> was his transfiguration. Solitary and almost incommunicado, he
> became the world's paramount mathematician.

He did not publish his ideas until about 20 years later. In the mean-
time he co-invented calculus. Among his notes were ideas that he later
expanded in a great work of three volumes, *Philosophiae Naturalis Mathe-
matica Principia*. Book 1 of the *Principia* deals with the laws of motion.
To apply kinematics to the motion of a material body, Newton introduced
two physical quantities: *force* and *mass*. He propounded three rules, known
today as *Newton's Laws*. Here we give them in modern phrasing.

4.1   Newton's Laws

What causes motion or a change of motion? This is a question that Newton
asked himself.

4.1.1   *The First Law*

What is a *law* of physics? It is assumed to be a truth that applies widely.

> *Law 1: A body at rest remains at rest, and a body in motion
> continues in motion in a straight line at constant speed, unless
> acted upon by an external force.*

So, what is a force? Simply stated, it is a push or a pull. This is what
causes a change of motion. In other words, an object keeps on going forever
in a straight line if no forces (e.g., friction) acts on it. Thus, friction must
be a force opposing motion.

4.1.2   *The Second Law*

> *Law 2: When a force acts on a body it causes an acceleration
> proportional to the force; the constant of proportionality is a
> measure of the inertia of the body, its mass.*

The first two laws are embodied in the vector equation

$$\vec{a} = \vec{F}/m\,, \tag{4.1}$$

where m is the mass of the body. The larger the mass, the larger the inertia
to a change of the velocity. Note that the direction of $\vec{a}$ is the same as that
of the force, $\vec{F}$. The definitions of force and mass are somewhat circular.
A force is something that tends to make an object accelerate; mass is the
property of the object that resists the force. Nevertheless they are useful
concepts; classical mechanics is built upon them.

In the metric system, the unit of force is the *Newton*, abbreviated as N, and the unit of mass is the *kilogram*, abbreviated kg. A force of 1.0 N acting on 1.0 kg causes it to accelerate at 1.0 m/s². For a gauge of magnitude, a penny has a mass of about 2.7 grams, i.e 2.7×10^{-3} kg.

If two or more forces act simultaneously on an object, it responds as if acted upon by their vector sum, as a single net force.

We need to distinguish between *internal* and *external* forces, An object may be composed of several pieces held together by internal forces. For example, all of the atoms of a solid body travel as a unit because there are forces between the atoms, which bind them together. These internal forces keep all of the parts of the body moving together when an external force acts on any part of the body. The internal forces cancel out, that is their vector sum vanishes, and it is only the *external* forces that determine the acceleration.

An object has *mass* wherever it is, even in space, far from Earth or other planets. But it has *weight* only when gravity pulls on it. If it rests on a scale it will make the scale deflect: *Weight is a force*. A body with mass m on Earth is acted on by the gravitational force of the earth **W**. If the body is free of all restraint, **W** causes it to fall with acceleration **g**, so according to Eq. (4.1),

$$\mathbf{W} = m\mathbf{g}. \qquad (4.2)$$

The gravitational force acts on the body even when it is not free. For example, if it hangs on a spring, the force **W** is resisted by the stretching of the spring; if it rests on a table, its weight is resisted by the rigidity (which actually involves a very slight compression) of the table.

Illustrative Example: A child begins pulling her sled on a patch of very slippery snow on the horizontal ground. The sled's mass is 2.0 kg, the cord is inclined at 45° with respect to the ground. The sled stays on the ground and is accelerating at the rate of 0.50 m/s². Calculate the total pulling force on the cord.

Answer: The force that is causing the acceleration is $\vec{F}_{\text{horizontal}} = m\vec{a} = (2 \times 0.5)$ kg m/s² $= 1.0$ N. This is the horizontal component of the total force; 1.0 N $= F_{\text{total}} \cos(45°)$. Since $\cos 45° = 0.707$, $F_{\text{total}} \approx 1.4$ N.

Illustrative Example: A crane is lifting a 500 kg piano by a cable that has a breaking strength S of 6000 N. What is the maximum upward acceleration

From a portrait by Kneller in 1689

Fig. 4.1   Portrait of Isaac Newton, courtesy Wikipedia.

that can be given to the piano through this cable? Assume standard gravity $g = 9.80$ m/s^2.

Answer: When the cable hoists the piano it must be strong enough to carry the piano's weight and also provide an upward acceleration; $S - W = ma$, $S = mg + ma$. Solving the equation for the acceleration, $a = (S/m) - g = 2.20$ m/s^2.

It is always useful to draw a force diagram. This is a disgram which shows all forces acting on a given object. In a complex problem, it is particularly helpful. We show a force diagram in Fig. 4.2.

Illustrative Example: Two masses, B of 2.00 kg and A with 20.0 kg are attached over a frictionless wheel, as shown in Fig. 4.2. There is a frictional force, $f = 18.0$ N, on block A with the table. Find the acceleration of the system, the net force on A and the tension in the rope joining the two bodies.

Solution: The forces on A are the tension of the rope pulling on it and the frictional force, all horizontal. There is a normal force **N** (upward

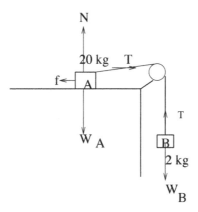

Fig. 4.2   Masses attached over a wheel.

from the table) and the weight of the block (downward), which cancel each other. The forces on B are the weight of the block and the tension opposing it. For the system as a whole, the forces are $\mathbf{W_B}$ and $\mathbf{f}$, so that $(2)9.8 - 18 = 1.6$ N $= (20 + 2)a$, and $a = 0.073$ m/s^2.

The net force on A $= \mathbf{T} - \mathbf{f} = $ T $-$ 18 N $= m_A a = (20$ kg$)(0.073) = 1.46$ N, so that T $= 19.5$ N. As a check, the acceleration of B (2 kg) is $a = (W - T)/m_B = (19.6 - 19.5)/2 = 0.05$ N, which is within the accuracy of the 3 significant figures. If you examine the individual bodies, we have for A

$$A : 2 \times 9.8 - T = 2a \tag{4.3}$$

$$B : T - 18 = 20a. \tag{4.4}$$

If you add these two equations, you find $2(9.8) - 18 = 22a$, which corresponds to that obtained for the system as a whole.

4.1.3   The Third Law

Law 3. For every action there is an equal and opposite reaction.

The Third Law means that an external force acting on a material body originates from some other body, which experiences an equal and opposite force. Note that the action and the reaction act on *different* bodies. Some examples will help make it clear.

Fig. 4.3   How to get off the ice.

Application of the Third Law shows that you can move the Earth! Just throw a ball into the air. You exert a force on the ball, which accelerates it upward. While you are in the act of throwing, the ball pushes downward on your arm. That force (the reaction) is transmitted through your body to the ground, which accelerates the Earth in the direction opposite to that of the ball. The acceleration is infinitesimal because the mass of the earth is huge compared to your mass and that of the ball.

Consider how a horse and wagon can travel. A horse is attached to a wagon. The horse and the wagon pull in opposite directions to the motion, but as long as the harness holds, they keep the same separation, and they move as a unit. The horse pushes against the ground, the resistance of the ground is equal and opposite; horse and wagon travel forward together, and the Earth recoils backward in response. Because of the earth's great mass, its motion is not detectable, but it moves, nevertheless. *Earth is an active partner of all of the travel that takes place on its surface.*

> *I shot an arrow into the air;*
> *it fell to earth I know not where,*
> *But this I know, and know full well:*
> *As arrow flew, Earth moved as well.*

4.2   Frames of Reference

When we apply Newton's Laws we normally consider measurements in our own *frame of reference*, the space where we are standing, as the *rest frame*.

But Newtonian dynamics holds in all frames of reference that are in a straight line, non-accelerated motion, relative to the rest frame. Such spaces are called *inertial frames*. For example, consider yourself traveling in a train at constant speed. You can toss a pencil up, and it falls back into your lap as normally as if you were sitting in a chair at home. When the pencil leaves your hand it is traveling forward just as fast as you and the train, and it keeps this forward motion thoughout its trajectory. The action is taking place in an inertial frame, and the dynamical behavior is the same as it would be if the train were stationary. If you examine the motion from the perspective of a person on the gound, then the pencil appears to travel in a parabola. This is due to the forward motion of the train superimposed on the vertical motion of the pencil in the train.

This experiment becomes more interesting when we consider dynamics in *non-inertial frames*, one where there is acceleration. If the train is accelerating while you toss the pencil into the air, it does not drop back into your lap; it lands somewhere behind you, toward the back of the train. It might seem as if a new force is operating, but it is just because the train's speed increased while the pencil was aloft. You can think of it as the train accelerating under the pencil.

4.3   Friction

Friction is so common that we are delighted/surprised when there is little or no friction: there's fun in sliding on ice, but it's a little less funny when we slip on a banana peel. We couldn't walk without the friction between our shoes and the ground, or drive cars without the friction between the tires and the road. All that is useful, but there are less desirable aspects to friction. The aging and breakdown of machinery is usually the result of the wear and tear of its rotating parts due to friction.

The microscopic mechanisms of friction are beyond the scope of introductory physics, but there are some rules for practical applications. In 1699 Guillaume Amontons, a French physicist, published a set of rules.

Sliding friction between solid bodies
(a) The friction between two sliding bodies depends on the state of both surfaces, such as moisture, roughness, cleanliness, and temperature.
(b) The force of friction is *parallel* to the surfaces in contact, and opposes sliding.

(c) Frictional force is proportional to the *normal force* N pressing the two surfaces together.

(d) The force of *kinetic friction* resists motion of two bodies *while sliding*:

$$F_k = \mu_k N , (4.5)$$

where μ_k is the *coefficient of kinetic friction* and N is the normal force. The coefficient of kinetic friction, μ_k, is approximately independent of sliding speed.

(e) The force of *static friction* resists motion of two bodies in stationary contact;

$$F_s \le \mu_s N , (4.6)$$

where μ_s is the coefficient of static friction. The $\le$ sign means that F_s resists a lateral force, with an equal and opposite frictional force, so that the net force $= 0$, *up to a maximum value*, $\mu_s N$. Once the object starts moving, it is μ_k that is relevant.

(f) μ_k is smaller than μ_s.

Rules (e) and (f) are simply demonstrated. For example, place a coin on a book, and gradually tilt the book at an increasing angle. The coin remains motionless, because static friction is equal and opposite to the downslope gravity until the tilt reaches a "critical angle". At the critical angle θ_c the limiting static friction is equal to the component of weight that tends to drive the coin down the slope. Thus, $\mu_s mg \cos \theta_c = mg \sin \theta_c$; therefore $\tan \theta_c = \mu_s$ As the tilt exceeds the critical angle the downslope force surpasses the limit of the force of static friction; then the coin suddenly begins to slide. The slide has an appreciable acceleration, because the downslope force is greater than the force of kinetic friction.

Illustrative Example: A 5 kg sled, with a 30 kg rider, is motionless on a 15° snowy slope. The coefficient of static friction is 0.30 and the coefficient of kinetic friction is 0.18.

(a) Draw a force diagram, indicating all forces acting on the sled.

(b) How large is the component of the weight parallel to the slope?

(c) How large is the frictional force, f?

(d) The sled is pushed to start it moving. How large is its acceleration?

Solution: (b) The downslope component of the weight is $mg \sin \theta = 0.259 \, mg = 88.8$ N.

The *maximum* static frictional force is $\mu_s mg \cos \theta = 0.29 \, mg$, which is

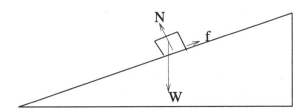

Fig. 4.4   Free force diagram of a box on a slope.

greater than the downslope gravitational component; therefore the actual static frictional force is just equal and opposite to the downslope gravitational force. Therefore the sled remains motionless.

(c) $F_s = 0.259 \sin \theta$.

(d) When sliding begins, the frictional force is kinetic friction, proportional to μ_k.

The sled accelerates due to the **net** downslope force, $F_{\text{net}} = mg \sin \theta - \mu_k mg \cos \theta$. Then $a = F_{\text{net}}/m = 0.85$ m/s^2.

4.3.1   Air resistance. Terminal velocity

Watch a bit of tissue paper fall. After accelerating a short distance, it falls at constant speed: it has reached its *terminal velocity*. Then crumple it up into a small ball, and let it drop again. It accelerates over a greater distance, and reaches a greater terminal velocity. The air resistance depends on its exposed surface area. More precisely, *the air resistance increases with an object's surface/volume ratio and also with its speed*. Speed dependence varies with the body's shape; the combination of speed and shape make for an extremely complicated force law. But at very low speeds the resistance is linear with speed.[a] For a body of arbitrary shape in some viscous medium such as air, the resistive force can be given by a general relation

$$F_{\text{res}} = -kv \,, \tag{4.7}$$

where k is a constant that depends on the shape and size of the body and the viscosity of the fluid; the units of k are Ns/m. A falling body is acted on by both gravity and air resistance, so the acceleration a is given by $ma = mg - kv$. Since $v = at$, we have

$$ma = mg - kat \,. \tag{4.8}$$

---

[a]For a spherical particle moving slowly in a viscous fluid of viscosity η, the resistive force is $F_s = 6\pi\eta a v$, where a is the radius and v is the speed.

Solving for the acceleration,

$$a = \frac{g}{1 + (k/m)t}. \tag{4.9}$$

Equation (4.9) shows that for short times, where $(k/m)t \ll 1$, the acceleration is approximately equal to its ideal free fall value g. For long times, where $(k/m)t \gg 1$, $a \to 0$. In that limit Eq. (4.9) yields the *terminal velocity* $v_t = at = mg/k$. The constant k depends on the density of the atmosphere, which is a function of the altitude.[b] The historical controversy over the speed and acceleration of falling objects was essentially an argument about the effects of air resistance. Galileo reasoned that the ideal, resistance-less motion could be deduced from the behavior of dense objects such as cannonballs, where air resistance had little effect. His opponents were unwilling or unable to imagine free fall in a vacuum.

Questions

Q4.1. When the car in which you are a passenger accelerates, you are pushed into the back of the seat rather than to the front. Why?

Q4.2. Consider two cars: Car A is strongly built, so that it can suffer a head-on crash with little damage. Car B is less sturdy; when it suffers a head-on crash the front end crumples. Which car is safer for its passengers? (Of course, there are several important factors in addition to the car's sturdiness, e.g. the strength of the bolts holding the seats to the frame, and the distance between the engine and the passenger compartment).

Q4.3. In a tug-of war, the two teams pull against each other; they are exerting equal and opposite forces on the rope. And even when one team is winning, by slowly dragging the other team, the forces are equal and opposite by Newton's 3rd law! Can you explain this?

---

[b]In 2008, a retired French army officer, Michel Fournier, planned to set an altitude record for free fall by jumping out of a helium balloon at an altitude of 25 miles. He would be dressed in an insulated and pressurized suit, with a sealed helmet fitted with an internal oxygen supply. His fall would not be completely free; a small parachute would keep him from spinning. According to calculations, his speed would accelerate to a terminal velocity of 663 mph at an altitude of 21.75 miles. Then as he reached denser and denser atmospheres, his terminal velocity would decrease: At 18.64 miles, 932 mph; at 15.53 miles, 663.9 mph; at 12.43 miles, 395.8 mph; at 6.21 miles, 192 mph. Total time of fall before opening his main parachute would be 8 minutes. Alas! Just before he could enter the capsule, the balloon was accidentally released, and it left without him!

Q4.4. What is the cause of a terminal speed? Arrange the following situations in order of the terminal speeds, with the slowest first.
(a) A sky diver, before the parachute opens
(b) A dime in water
(c) A feather in air
(d) A penny in honey

Q4.5. Why is the the acceleration of a heavy falling object the same as that of a light one, since the weight of the heavier one is much larger?

Q4.6. A 4 kg block has an acceleration that is twice that of an 8 kg block. Compare the forces.

Q4.7. Two forces of the same magnitude act on an object. The first one is due East and the second one is due North. What is the direction of the acceleration of the object?

Q4.8. A block on a table has two forces acting on it. The first is its weight and the second one is the normal force in the opposite direction. Are these forces action and reaction? If not, what are the reaction forces?

Q4.9. A ball is suspended from the ceiling by a string. What are the forces on the ball? What are the reaction forces?

Q4.10. Two blocks of the same mass are tied together by a rope. They move in unison on a plane surface at constant velocity. Consider two cases: (a) a rope is tied to one block and pulls both blocks. (b) The plane is inclined and the blocks slide down its surface, which has friction. Is the tension in the rope = 0? Explain.

Q4.11. An elevator is accelerating downward. Is the normal force on an individual in the elevator larger than, equal to, or smaller than the weight of the individual when the elevator is stationary? Explain.

Q4.12. The cable on an elevator snaps. Is the force of gravity on an individual in the elevator = 0? What would a scale the individual stands on read? Explain.

Problems

P4.1. A car is disabled by a malfunctioning engine, so a tow car has been summoned. The tow cable has a breaking strength of 1.5×10^2 N; the car has a mass of 7.0×10^2 kg. Assuming that the disabled car can roll without friction, what is the least time that it can be towed a distance of 100 m?

P4.2. A student is bringing 20.0 kg of books back to the library. She gets into an elevator at the ground floor, pushes the button for the 6th floor, and the elevator starts with an acceleration of 0.50 Gals. What is the weight of the books (e.g., as measured on a scale) during the acceleration?

P4.3. A student whose mass is 70.0 kg enters an elevator and steps on a scale. The elevator is ascending at constant speed, when the cable breaks and the cabin falls. Air resistance slows the fall, and 3.0 seconds after the break the cabin falls at its constant terminal speed of 15.0 m/s. 10.0 seconds after the break, the cabin hits the bottom of the shaft.
(a) One millisecond before the cable breaks, what is the student's weight (the scale reading)?
(b) One millisecond after the cable breaks, what is the student's weight?
(c) 6.0 seconds after the break, what is the student's weight and what is k?

P4.4. A block of wood rests on a ramp inclined at an angle θ from the horizontal. The block's mass is 5.0 kg; its static coefficient of friction against the ramp is 0.30 and its kinetic coefficient is 0.20.
(a) What is the maximum angle θ_c at which the block can remain at rest?
(b) If the angle slightly exceeds θ_c, what is the acceleration of the block down the ramp?

P4.5. A 5.0 kg stone falls onto a glacier, whose surface is a locally flat plane inclined at an angle 25° from horizontal. The static coefficient of friction between the stone and the glacier is 0.30 and the kinetic coefficient is 0.10.
(a) Is the static friction strong enough to keep the stone from sliding downhill?
(b) If the stone begins to slide, what will be its acceleration?

P4.6. A pencil is dropped from a height of 1.0 meter above the floor, in an airplane that is:
(a) traveling horizontally at constant speed,
(b) accelerating horizontally at 3.0 m/s^2,
(c) accelerating horizontally at 3.0 m/s^2 while rising at a constant speed of 0.50 m/s.
In each case, where does the pencil hit the floor?

P4.7. A pencil is dropped from a height of 1.0 meter above the floor, in an airplane that is accelerating before takeoff at 2.0 m/s^2. Where does the pencil hit the floor?

P4.8. Challenge. A mop is being pushed along a floor. The mop head weighs 10.0 N, and the mop handle (of negligible weight) is inclined at 20° from vertical. The coefficient of kinetic friction between the mop and the floor is 0.20. How large a force, exerted downward along the handle, is needed to push the mop? Hint: the pushing force has a component that adds to the weight. Make a careful sketch.

P4.9. A girl slides down a slick snow bank that makes an angle of 7.8° with the horizontal. Her speed is 1.6 m/s. The bank levels off at the bottom to a flat area. The coefficient of kinetic friction does not change. How much time does it take her to come to a stop on the flat part?

P4.10. The driver of a car of mass 1500 kg traveling at 90.0 km/hr suddenly applies the brakes. The coefficients of friction between the wheels and the road are $\mu_R = 0.30$, $\mu_s = 0.70$ and $\mu_k = 0.20$. What is the minimum force between the wheels and the road that will cause the tires to skid? Hint: The car is then sliding rather than rolling.

P4.11. Two weights of 20.0 N and 10.0 N are tied together by a rope which passes over a frictionless wheel. Find the acceleration of the system and the tension in the rope.

P4.12. Two blocks, A (5.0 kg) and B (8.0 kg) are resting on the floor and are tied together by a rope, The two masses are accelerated by a pull of 20.0 N exerted on block A. The kinetic frictional force on A is 5.0 N and that on B is 8.0 N.
(a) Draw a force diagram.
(b) Find the acceleration of the system.
(c) Determine the tension in the rope tying the two blocks together.

Answers to odd-numbered questions

Q1. It is the relative motion that makes it appear that you are pushing into the seat back. From the standpoint of a stationary observer, the car's seat back is accelerating forward, and it pushes to accelerate you.

Q3. The two teams pull equally against each other. Both dig in their heels, and push against the ground, and *the winning team pushes harder against the ground.* Many students have found this problem one of the most puzzling in the course.

Q5. Because the mass increase is proportional to the weight.

Q7. The acceleration is NE at $45°$ from the East.

Q9. There is the weight of the ball and the pull of the rope. The reaction forces are the force on the earth and the tension in the rope.

Q11. The weight is larger than the normal force since the acceleration is down.

Answers to odd-numbered problems

P1. The breaking strength S limits the acceleration to: $a_{max} = S/m = 0.24$ m/s^2. Relating the distance x to the time by $x = (1/2)a_{max}t^2$, we find $t = 30.6$ s to pull the car 100 m. In this time, with the maximum acceleration, the final velocity would be 6.54 m/s = 23.6 km/hr.

P3. (a) 70.0 kg   (b) 0   (c) 70.0 kg; $k = 45.7$ Ns/m.

P5. (a) The downslope force is $mg \sin 45°$. It is opposed by the force of static friction $\mu_s mg \cos 45°$, which is insufficient to keep the stone from sliding. It must be supplemented by a force upramp $F = mg \sin 45° - \mu_s mg \cos 45° = (1 - \mu_s)(0.707)$ mg = 0.495 mg $\approx$ 0.50 mg.
(b) To move the stone upramp at constant speed requires a force $F = mg \sin 45° + \mu_k mg \cos 45° = 0.778$ mg $\approx$ 0.78 mg.

P7. At a distance forward, $x = \frac{1}{2}at^2$, where $a = 2m/s^2$; t is the time to fall, $t = \sqrt{2y/g}$, where $y = 1$ m; $t = 0.45$ s; thus $x = 0.20$ m.

P9. $\mu_k = \tan(7.8°) = 0.137$ for a constant speed down the slope. At the bottom, the acceleration is $a = -mg\mu_k/m = -1.34$ m/s^2; since $(v_f - v_i)/a = t$, we find $t = -1.4$ m/s$/(1.34$ m/s$^2) = 1.04$ s.

P11. The difference of the weights is 10 N, so that 10 N $= (m_A + m_B)a = (30/9.8$ kg$)a$ and $a = 3.3$ m/s^2. The tension T accelerates the 10 N weight, $T - 10 = (10/9.8)a = 1.02\ a = 3.4$ N, and $T = 13.4\ N = 13$ N. Check: $(20 - T) = (20/9.8)a = (2.04)(3.3$ m/s$) = 6.73.\ T = 13.3.$ To the desired accuracy, $T = 13$ N.

Chapter 5

Work and Energy

5.1 The Quest for Perpetual Motion

The search for a perpetual motion machine has a long history. The aim was to find or create a device that could continue to move forever, and while moving perform useful work. (In fact, friction alone would make such a machine impossible, even without performing useful work). It was one of the two great and impossible quests of medieval technology; the other was the Philosopher's Stone, that would turn base metals into gold.

A perpetual motion machine would violate the Law of Conservation of Energy, which was discovered by experiment in the 19th century. Nevertheless, the U.S. Patent Office continued to receive such applications. In fact, it was because the Patent Office had to analyze so many proposals for perpetual motion that in 1911 it instituted a new rule: every patent application for a perpetual motion machine had to be accompanied by a working model that could run for one year. But this very sensible rule was abandoned, after a few years, for unexplained reasons.

5.2 Mechanical Work

We are quite familiar with "work"; a good example is *homework*. Although you may have a lot of it, it is not *work* in Physics. The word *work* has an exact meaning in physics. Mechanical work requires the application of a force to an object and the displacement of that object. The work done on an object is the product of the applied force times the distance the object travels *if the force is applied in the direction of motion*. If the force is applied in some other direction, it is only the component of the force in the direction of motion that matters. An example is shown in Fig. 5.1.

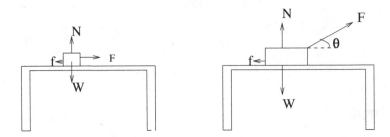

Fig. 5.1   Forces acting on a box being pulled by a force $\vec{F}$. The frictional force is labeled f.

When we lift a weight, at a slow but constant speed, for most of the way, our effort depends on both the weight and how high we lift it. It is the product of the two that strains us; twice the weight means twice the effort; twice the weight *and* twice the distance takes four times the effort. The amount of effort is a measure of the *mechanical work* ΔW.

$$\Delta W = F_{\parallel} d = dF \cos \theta , \qquad (5.1)$$

where $F_{\parallel}$ is the force exerted parallel to the distance d, and θ is the angle between $\mathbf{F}$ and $\mathbf{d}$. (We will use ΔW to distinguish *Work* from *Weight*, which we denote by $\mathbf{W}$). The unit of work is the *Joule* and 1 J = 1 N times 1 m.

The condition that the force and the distance, or a component of the force, be parallel is essential. Work can be positive or negative. If the force is applied opposite to the direction of motion, then the object slows down and the work done on the object is negative. In contrast, if the force and the distance are perpendicular, we find that $\Delta W = 0$, since $\cos \theta = 0$. Thus, if we just carry a weight horizontally, we do no mechanical work. It may be very tiring to carry it, but that's a matter of how our muscles act; according to our definition, it's not mechanical work. Just to make sure that we don't confuse the two, we call ΔW *external work*. Ideally, we could be replaced by a rolling cart, which would need no force to move the weight at constant speed.

Illustrative Problem: Analyze the work being done by pulling a block along a table at constant velocity. Show all forces and the work they do when the force is not necessarily parallel to the table.

Solution: The forces are shown in Fig. 5.1 for both a pull along the table and one that is in another direction. The weight (W) and the normal force

F

W

Fig. 5.2   Work done in lifting a box of weight $\vec{W}$. The force $\vec{F}$ is the applied force.

(N) do no work. The work done by the frictional force (f) is $-fd$. The work done by the pulling force is Fd in the first instance and $Fd\cos\theta$ in the second case.

Both work and energy are scalars; that is one advantage of using them rather than Newton's laws.

5.3   Potential Energy

What is the physical consequence of raising a weight; is there anything different about the weight when it has been raised? The answer is Yes: *It has the potential for a variety of actions that it did not have before we raised it.* For example, if it fell, it could smash something on the ground; maybe usefully, like crushing grapes, or mashing potatoes. See, for instance, Fig. 5.4. Or we might rig up some sort of rope and pulley arrangement, so that it could operate a machine while descending gradually. (Some old clocks work on this principle). The details aren't important; what matters here is that, when raised, the weight has the potential for producing some external work. We have increased its *potential energy, U*:

$$\Delta U = \Delta W .\qquad (5.2)$$

We use the symbol U for potential energy. Changes in potential energy are independent of the paths that brought about the change, as long as friction can be neglected. As an example, we compare the work done in lifting a weight straight up, with the work done in moving the weight up an inclined

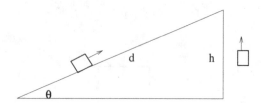

Fig. 5.3   Comparison of moving a block up an incline versus straight up.

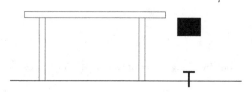

Fig. 5.4   Potential energy can be used to do work, such as hammering a nail into the floor.

(frictionless) ramp (Fig. 5.3), to reach the same height. In the first case, we apply a vertical force mg, and move it a vertical distance h, so

$$\Delta W = \Delta U = mgh\,. \qquad (5.3)$$

Now let us raise the weight by pushing it up an inclined ramp. The force we have to apply is smaller, $mg\sin\theta$, but the distance is greater. As shown in Fig. 5.3, the distance is $d = h/\sin\theta$. So when we take the product of force and parallel distance the angle factors cancel, and we get $\Delta W = mgh$, the same result as moving it straight up. In fact, we could move the weight over any complicated path, with ups and downs, but if it ended up at an increased height h, the net work would be mgh, the same as in a simple, straight lift, assuming that the path doesn't involve friction, of course. Because the work is done against gravity, the potential energy is often called *gravitational potential energy*. The unit of energy is clearly the same as that of work; in the MKS (meter-kg-s) system it is the Joule.

5.4   Kinetic Energy

If we apply a constant horizontal force, **F** to a mass m at rest on a table, and neglect friction, the acceleration is $\vec{a} = \vec{F}/m$, and if the force operates over a distance d parallel to the force, the work done is $Fd = mad$. If the force acts for a time t, we relate the time to the distance by $d = \frac{1}{2}at^2$. Expressing the acceleration in terms of the final velocity v, $a = v/t$ and eliminating the time, we get a relation for the *kinetic energy* K:

$$\Delta W = K = \frac{1}{2}mv^2. \qquad (5.4)$$

The same thing can be done vertically for a falling mass. If the mass falls freely, starting from rest, its potential energy turns into the energy of motion: *kinetic energy*. The work done by gravity in moving the object a distance d is again $Fd = (ma)d = \frac{1}{2}mv^2 =$ gain in K. Although we derived K for specific processes, Eq. (5.4) is the general expression for kinetic energy; it does not depend on how the object acquired its speed. And it does not depend on direction: Work and kinetic energy are scalars. There is one proviso: for Eq. (5.4) to be valid, the speed must be much smaller than the speed of light, $c = 3 \times 10^8$ m/s. When $v \to c$, most of the equations we have given are altered.

Illustrative Example: A 5 kg mass is lifted to a height of 1 m above ground. It then falls a distance of 80 cm where it encounters a peg.
(a) How much did the P.E. of the mass change during the lifting process?
(b) What is the kinetic energy of the mass just before it hits the peg?
(c) What is the potential energy of the mass relative to ground ($=0$)?
(d) What is the energy imparted to the peg?

Solution: (a) $\Delta U = mgh = 5$ kg $\times$ 9.8 m/s^2 $\times$ 1 m $= 49$ J
(b) $K = |\Delta U| = mg(h - h') = 5$ kg $\times$ 9.8 m/s^2 $\times$ 0.8 m $= 39$ J
(c) $U = mgh' = 5$ kg $\times$ 9.8 m/s^2 $\times$ 0.2 m $= 9.8$ J
(d) The energy imparted to the peg is the kinetic energy of the mass: 39 J. The mass retains the extra P.E. (9.8 J).

5.4.1   Non-constant forces

So far, we have assumed constant forces. What happens when the applied force is not constant? The problem gets more complicated, except if the force increases (or decreases) linearly. In that case, we can take the average force, which is $\frac{1}{2}(\vec{F}_i + \vec{F}_f)$, where i and f stand for initial and final. This is familiar from our study of accelerated motion, where the average velocity is similarly related to the initial and final velocities. In all cases $\Delta W = \Delta K$ (or ΔU).

5.5   Power

Power is the *rate* of doing work, the rate of energy usage:

$$P = \Delta W / \Delta t. \qquad (5.5)$$

Its symbol is P, and its unit, in the metric system, is the familiar *Watt*: 1 Watt = 1 Joule/second. Another familiar term is "horsepower" hp; the conversion factor is 1 hp = 746 watts. (A horse couldn't last long working at the rate of a horsepower)!

If a constant force in the direction of motion is moving an object for a distance d parallel to the force, at a constant speed v (against gravity, or frictional resistance, for example), the power output is the product of force and speed:

$$\Delta W = Fd; \quad P = \Delta W / \Delta t = Fd / \Delta t = Fv \qquad (5.6)$$

Illustrative Example: You (mass $m = 60$ kg) are walking up a 20° incline (a large slope!) at a speed of 0.5 m/s. Neglect friction. How much power are you expending if friction can be neglected?

Answer: The force you exert to counter the downslope gravity is $F = mg \sin \theta$. If you move at the speed v, you are expending power $P = Fv = vmg \sin \theta$. With the values given, $P = 101W \approx 100W$.

5.6   Conservation of Energy

Even beyond the fact that it is a physical law, it seems that conservation of energy makes sense. People say things like, "You can't get something for nothing" and "There's no free lunch" when someone comes up with another crazy idea like running a car on water instead of gasoline. But if you want to establish a rule as a physical law, you need firm evidence, either an experimental test or a theoretical proof. The Law of Conservation of Energy has both.

The conservation of total energy is one of the conservation laws of physics. Like all conservation laws, it follows from a *symmetry principle*, as shown by the famous woman mathematician (when women were still very scarce in the sciences) Emmy Noether.

The story of Amalie (Emmy) Noether is an interesting one, as it illustrates the difficulties that women still had at the beginning of the 20th century (also later). She was born in 1882 in Aligner, Germany. She first studied French and English so that she could teach these subjects, but never did so. After high school, she became interested in mathematics. Both her father and brother were mathematicians. The University of Aligner refused to admit her (she was a woman!), but at the behest of her father, a professor of mathematics there, allowed her to sit in on courses. After two years, she passed the exam that would allow her to become a doctoral student, and the university finally admitted her. She wanted to teach mathematics, but was not allowed to do so. She did teach her father's classes when he was sick or absent. Of course, she was not paid. She was invited to the University of Goettingen by two famous mathematics professors there. They had to fight their colleagues for four years before she got a position with a small salary. It was at the beginning of her stay here that she proved the theorem that conservation laws are related to symmetries. Emmy was a pacifist and hated WWI. When Hitler came, she promptly lost her position, but was famous enough by then to obtain a position at Bryn Mawr, where she continued teaching until her death in 1935.

The symmetry that results in the conservation of energy is that under time translation. The laws of physics are invariant under a time translation. That is, you can do an experiment today, tomorrow, or the day after, and you should get the same result. Mathematically, the connection of time translation to energy conservation follows from the Lagrangian theory for motion, which is beyond us.

5.6.1 *Experimental tests of the law of conservation of energy; Other forms of energy*

There is a great amount of experimental evidence for the conservation of total energy. For example, it can be converted from one type to another. Mechanical energy can be converted into heat, as in friction, but is conserved in the process. In fact, the unit of energy, the *Joule* comes from an experiment by the physicist James Prescott Joule, who showed this conversion; see Ch. 10. Light is another form of energy. The sun supplies us with this form of energy and is also responsible for photosynthesis, the conversion of this energy into food for plants. We shall see other instances of energy conservation in later chapters.

A pendulum is a very simple machine that cycles energy between potential and kinetic. At the end of each swing the pendulum bob is momentarily stationary, at a height H above the bottom and its energy is

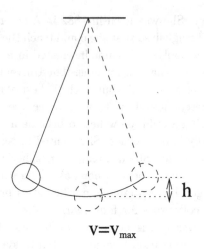

$$V = V_{max}$$

Fig. 5.5   A simple pendulum.

purely potential energy, if the P.E. is taken as zero at the bottom of the swing. $\Delta U = U = mgH$. When the bob returns to the bottom of the swing the potential energy has been converted completely into kinetic energy: $K = \frac{1}{2}mv_{max}^2$. (We are, of course, neglecting air resistance). At intermediate positions, say at a height h above the bottom, the energy is a sum of potential and kinetic energy, and the sum or total energy is a constant. At the height h,

$$E(total) = mgh + \frac{1}{2}mv^2 = mgH . \qquad (5.7)$$

A pendulum's motion is a simple example of the *Law of Conservation of Energy (LCE)*. By a combination of experiment and analysis, the principle has become broadened, to apply to all forms of energy. The kinetic and potential energy of a pendulum, and the *elastic* (as contrasted to *gravitational*) potential energy of a spring are types of *mechanical energy*. Other forms of energy are: chemical, thermal, electrical, magnetic, (together, they are electromagnetic), and nuclear. There are various processes that can convert from one form to another, but whether converted or not, the total energy of a *closed system* is conserved. A couple of examples illustrate the wide scope of the LCE.

Imagine a spring, wound up and then fitted into a glass jar. Now the jar is filled with an acid which completely dissolves the spring. What happened to the compression energy of the spring? We do not know if such an experiment was ever done, but we can rely on the LCE to predict that the heat evolved by the reaction would be greater than if the spring had not been compressed, and the excess heat would be equivalent to the energy of compression.

In nuclear reactions, the total amount of energy is also conserved. But in one particular nuclear reaction (the emission of electrons or so-called *beta decays*), careful bookkeeping indicated a problem: repeated measurements of the masses and energies showed a loss. Some well-known physicists were willing to give up energy conservation. Out of a firm belief in the conservation of energy, Wolfgang Pauli suggested that the missing mass/energy was actually carried away by a particle, a particle so hard to detect (no charge and no mass) that it had escaped notice. Enrico Fermi named the imaginary particle *neutrino*, "little neutral one". More than 30 years later, experiments using massive detectors confirmed the existence of neutrinos; they are in fact abundant, but very difficult to detect.

5.6.2 *"Energy conservation" is not conservation of energy*

A physics student can raise an eyebrow at the campaign for "Energy Conservation". *Of course*, the student thinks, the laws of physics don't need any special efforts from government, car manufacturers, or citizens. Energy is conserved, whether it is in useful forms, or converted into heat, and degenerated into toxic chemicals. We cheer the campaign for "energy conservation", but the Law of Conservation of Energy is obeyed, even when the energy is "wasted".

5.7 Simple Machines

There are many simple machines, the purpose of which is to help you exert a smaller force, $\vec{f}$, than you would have to supply if you applied the force, $\vec{F}$, directly to do some work

$$mechanical\ advantage = \frac{|\vec{F}|}{|\vec{f}|} \tag{5.8}$$

A particularly simple machine is the lever (Fig. 5.6), which helps you to raise a heavy object. This simple machine, like all others, does not save you work, but provides a mechanical advantage: the ratio of the weight lifted **W** to the force applied **F**. The work done is the change of P.E., ΔU; the

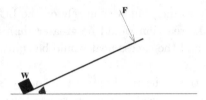

Fig. 5.6   A simple lever to help raise a heavy object.

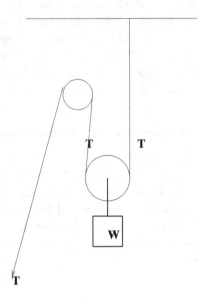

Fig. 5.7   A simple pulley system with $2\mathbf{T} = \mathbf{W}$ to help raise the heavy object of weight **W**.

force must be applied over a larger distance than the weight is lifted (by the ratio of the mechanical advantage).

Another simple machine is a system of pulleys. Depending on the arrangement, the mechanical advantage may be a factor of 2, as shown in Fig. 5.7, or larger, depending on how many ropes hold up the weight **W**. In the case shown, with two ropes, each rope (tension) only lifts half the weight, but the person pulling on the rope pulls it twice as far as the weight is lifted.

Questions

Q5.1. What happens to a car's kinetic energy when the driver applies the brakes?

Q5.2. Why is a car's average mileage (miles/gallon) better in highway driving than in city driving?

Q5.3. It has been proposed that a national speed limit be imposed (55 mph is suggested) as an energy conservation measure. Why would it be an energy conservation measure?

Q5.4. What are the various forms of energy that are involved when you jump on a trampoline?

Q5.5. Where does the energy come from when you "pump up" a playground swing while you are in it?

Q5.6. Why does rubbing your hands warm them?

Q5.7. A car's energy comes from the burning of gasoline in its engine. Where does the gasoline's energy come from?

Q5.8. Does the frictional force do work on an object? Does the normal force (N) do work on an object moving down an incline? Explain.

Q5.9. If there is a single force on an object, does its work necessarily increase its K.E.? Explain.

Q5.10. The work done on a block increases its velocity by a factor of 2. Is the energy increased by the same factor? Explain.

Q5.11. The work done in lifting an object may not increase its velocity. Explain.

Q5.12. If you compress a spring, are you doing work on the spring? Does the energy of the spring increase? Explain.

Q5.13. If a pole vaulter acquires a greater velocity prior to using the pole to reach a high bar, does she increase the height she can reach? Explain.

Q5.14. The mechanical advantage of a lever is 3. In order to lift a weight a distance of 10 cm. Over what distance must the force be applied?

Q5.15. For Q5.14 compare the power of lifting the weight with and without the lever, if the time taken in both cases is the same.

Q5.16. In the pendulum used in the text, what is the work transmitted by the string that holds up the bob during 1/2 cycle?

Q5.17. When a weight lifter lowers the weight, does he do positive or negative work? Explain.

Problems

P5.1. A 70.0 kg skier, starting from rest, descended a 1.0 km-long slope inclined at 15° from horizontal, and reached a speed of 10.0 m/s at the bottom. How much energy was lost to friction?

P5.2. A waterfall's rate of flow is 10.0 kg/sec, and it falls from a height of 10.0 m. If all of the water's change of potential energy is converted to electrical power, how many watts will it produce?

P5.3. A 2.0 kg block slides down a frictionless ramp. The ramp is 10.0 m long, inclined at an angle of 30°. Using the kinematic equations in Ch. 2.
(a) Calculate the speed of the block at the bottom of the ramp, and the corresponding change of potential energy.
(b) If, instead of sliding, the block had fallen from the top of the ramp, what speed would it have just before hitting the ground, and what would be the corresponding change of potential energy?
(c) Do the potential energy changes in (a) and (b) depend on the path of the motion?

P5.4. A catapult is "loaded" when its throwing arm is pulled back by a steady force of 100.0 N for a distance of 3.0 meters. Using the law of conservation of energy how high can it throw a 5.0 kg stone?

P5.5. How much external work do you expend when you drag a 30.0 kg sled up a 15° hill a distance of 50.0 meters?
(a) Assume zero friction.
(b) Assume a friction coefficient $\mu_k = 0.20$.
(c) If you do this in 4.0 minutes, what power did you exert?

P5.6. How much external work do you expend when you carry a suitcase weighing 50 pounds:

(a) 1 km on a horizontal sidewalk?

(b) while standing on an escalator that rises 10.0 meters?

P5.7. A bowling ball of mass 5.0 kg at the end of a long cord is lifted to a height of 25 cm above the bottom and released. What is its speed as it crosses the bottom of its path?

P5.8. A pendulum with a bob of mass 500.0 g is lifted 25.0 cm and released.

(a) What is the bob's speed when it crosses the bottom?

(b) At what height is the speed of the bob half of its maximum? At that point what are the K.E. and P.E.?

P5.9. A lever is used to lift a 50.0 kg weight a distance of 10.0 cm in 2.0 s. Find the work and power expended by the user of the lever.

P.5.10. For the pulley illustrated in Fig. 5.7, what force must be applied to lift a weight of 10.0 kg? Over what distance must the rope be pulled to lift it 25 cm? If the lifting is done in 30.0 s, what is the average power expended?

P5.11. A ball of mass 14 g is thrown straight up at an initial speed of 12 m/s.

(a) Use energy conservation to determine the height to which it rises.

(b) Determine the work done by gravity as the ball rises the first meter and the last meter of its upward path.

(c) Determine the work done by gravity as the ball falls 1 m from its highest point.

(d) What is the ball's K when it reaches a height of 3.0 m? What is its U at this location if $U = 0$ at the start?

P5.12. (more difficult) A young woman uses a sturdy tree limb jutting out from a tree on a small piece of land near the shore to tie a rope to act as a swing to jump into a lake. She stands on a ledge that is 7.0 m above the water line and begins her swing with zero velocity.

The length of the rope to the point of attachment of the woman is 6.1 m. Consider the entire mass of 62 kg of the woman to be concentrated at the point of attachment. She swings out from shore to above the water and lets go close to where the rope reaches its lowest point. (See Fig. 5.8).

(a) Does it make any difference to the speed with which she hits the water whether she lets go on the swings downward or upward motion?

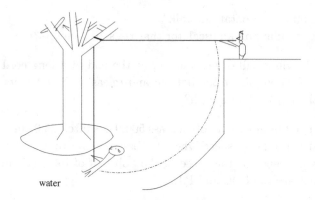

Fig. 5.8   Swing for woman above lake.

(b) What is the speed with which she enters the water?
(c) How far from shore will she land if she lets go at the bottom of the swing and the attachment to the limb is 2.0 m from shore?

P5.13. A skier of mass 83 kg descends a slope. The kinetic coefficient of friction is 0.11. The angle of the slope is 20°. (The component of $\vec{W}$ along the slope is $|\vec{W}| \sin \theta$).
(a) Use energy conservation to determine the skier's speed 30 m down (along) the slope if she starts from rest.
(b) What height, h, of a bump or hill can she go over at that height? Assume that h is the vertical height of the hill. That is, how high a vertical rise can she go over? Neglect friction on the bump.

P5.14. Two pendula have the same length of rope. One bob has mass $m = 45$ g and the other one a larger mass $M = 1.2$ kg. They are both lifted to the same height, 0.82 m, before being let go. Neglect friction.
(a) Which bob has a larger kinetic energy at the bottom of its swing? Why?
(b) Find the ratio of the maximum velocities of the two bobs.
(c) Which one, if either, would you expect to get to the bottom first? Why?

P5.15. What is the mechanical advantage of an inclined plane making an angle $\theta = 30°$ with the horizontal?

P5.16. (more difficult) A 3.1×10^3 kg satellite is sent aloft to get close-up pictures of the moon. Its radius around the moon is 3.1×10^6 m. The mass of the moon is 7.38×10^{22} kg. Neglect the effect of the earth. It is desired to move the satellite closer to the moon. $G = 6.67 \times 10^{-11} \text{N} - \text{m}^2/\text{kg}^2$.

(a) Must the satellite's speed increase or decrease? Why?

(b) What force is needed to get the satellite to circumnavigate the moon at a radius of 2.7×10^6 m from its orbit at 3.1×10^6 m?

(c) How much work is needed?

(d) What will be the speed of the satellite at this new radius?

Answers to odd-numbered questions

Q1. It is converted to heat.

Q3. Less kinetic energy is lost whenever the car stops. Also, die resistance increases with speed.

Q5. The energy comes from your kinetic energy, as you move your body back and forth. You are moving your center of mass in resonance with the natural frequency of the swing (with you aboard).

Q7. The energy comes from the chemical combination of gasoline and oxygen... the burning of gasoline. Ultimately, petroleum's energy came from the Sun, which caused the algae and all the other petroleum's source plants to grow.

Q9. The force may decrease the K.E. (as in slowing a car).

Q11. The work may increase the potential energy and not the K.E.

Q13. The additional energy should increase the height reached.

Q15. The power is the same.

Q17. From your muscles.

Answers to odd-numbered problems

P1. The skier's potential energy at the top of the hill was mgH. The skier's kinetic energy at the bottom of the hill was $\frac{1}{2}mv^2$. The difference between the two values is the energy "lost" to friction. $E_{lost} = mgH - (1/2)mv^2 = 70[9.8 \times 10^3 \sin 15° - \frac{1}{2}100] = 1.78 \times 10^5$ J.

P3. (a) $\Delta(P.E.) = mgl \sin 30° = mgh = 98$ J.
(b) $v = \sqrt{2gh} = 9.9$ m/s. (c) No.

P5. (a) $\Delta W = MgL \sin \theta = 3.81 \times 10^3$ J.
(b) $\Delta W = MgL \sin \theta + \mu_k Lmg \cos \theta = 3.81 \times 10^3 + 2.84 \times 10^3$ J $= 6.6 \times 10^3$ J.
(c) 28 W.

P7. The $\Delta P.E. = 5$ kg $\times 9.8$ m/s$^2 \times 0.25$ m $= 12.3$ J. This is the K.E. at the bottom $= (1/2)mv^2$. Thus $v = \sqrt{2E/m} = \sqrt{24.6/5} = 2.2$ m/s.

P9. The $\Delta W = (50 \text{ kg})(9.8 \text{ m/s}^2)(0.1 \text{ m}) = 49 \text{ J}$, $P = \Delta E / \Delta t = 49 J/2s = 25 \text{ W}$.

P11. (a) $mgh = (1/2)MN^2$, or $h = v^2/(2g) = 7.3 \text{ m}$.
(b) $\Delta W = -mg\Delta h = -0.14 \text{ J}$.
(c) $\Delta W = mgh = 0.14 \text{ J}$.
(d) $\Delta U = mgh = 0.41 \text{ J} = -\Delta K$, $K_i(1/2)mv^2 = 1.0 \text{ J}$. So, K at 3.0 m is 1.0 J -0.41 J $= 0.59$ J and $\Delta U = 0.41$ J.

P13. (a) Initially the energy is mgh. 30 m down the slope $\Delta U = mg\Delta h$ and $K = (1/2)mv^2$ – the energy lost by friction, $-\mu_k N\Delta L = \mu_k mg \cos\theta\Delta L$. So, $v^2 = 2g\Delta L(\sin\theta - \mu_k \cos\theta) = 161 \text{ (m/s)}^2$ and $v = 12.7$ or 13 m/s.
(b) $h = v^2/(2g) = 8.2 \text{ m}$.

P15. $1/\sin\theta = 2.0$.

Chapter 6

Impulse and Momentum

6.1 Impulse

Momentum has a very definite meaning in physics, which differs from that in everyday use. We begin by introducing *impulse*, which is related to momentum.

When we think of an impulse we may think of a kick or a rapid signal. Indeed, even in physics, *impulse* is used primarily when a large force is applied for a short time. An example is a golf driver and the golf ball it hits, or a tennis racquet and a tennis ball. In both cases the applied forces are large, but the contact time is short. It may be difficult to measure these forces and the time of contact. Furthermore, the force tends to increase and then decrease; it is not constant; see Fig. 7.1. We can, however, replace it with an average constant force over the time of contact (called $\Delta \tau$ in Fig. 7.1). The *impulse*, $\vec{I}$ is the average force $\vec{F}$ times the time of contact

$$\vec{I} = \vec{F} \Delta t \,. \tag{6.1}$$

We can therefore write

$$\vec{F} = m\vec{a} = m\frac{\Delta \vec{v}}{\Delta t}\,, \tag{6.2}$$

$$\vec{I} = \vec{F} \Delta t = m\Delta \vec{v} = m(\vec{v}_f - \vec{v}_i)\,. \tag{6.3}$$

Note that the impulse depends on both the force and the time of application.

If the force is in the direction of motion, then the work done on the object by the average force during contact is this force times the distance the object moves during the application of the force,

$$W = \vec{F}d \,. \tag{6.4}$$

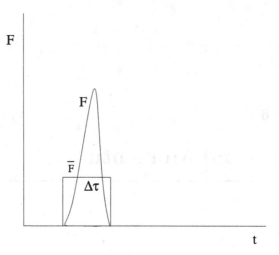

Fig. 6.1   Typical time variation for an "impulsive" force. The box is the average force.

6.2   Momentum

We define the *momentum* as

$$\vec{p} = m\vec{v}. \tag{6.5}$$

The units of momentum are kg-m/s. From Eq. (6.3), we see that

$$\vec{I} = \Delta\vec{p}. \tag{6.6}$$

Note that a small mass may have the same momentum as a much larger one if its velocity is much larger. We know that $\vec{F} = m\vec{a} = m\frac{\Delta\vec{v}}{\Delta t}$. If we multiply both sides of the equation by Δt, the left hand side of the equality is seen to just be the impulse $\vec{I}$, and the right hand side is the change of momentum.

Illustrative Problem: Rafael Nadal's tennis racquet exerts an average force of 500 N when it hits a tennis ball at service; he gives it a horizontal velocity of 52 m/s when he serves. The time of contact of the ball with the tennis rout is approximately 0.001 s.
(a) Determine the magnitude of the impulse.
(b) Determine the mass of the tennis ball.
(c) What was the work done by Natalie if the tennis raquet moved 90 cm horizontally during the serve?

Solution: (a) $\vec{I} = \vec{F}\Delta t = 500$ N $\times 0.001$ s $= 0.5$ Ns.

(b) The ball had zero horizontal velocity at the start and 52 m/s at the end. $\Delta|\vec{p}| = 52$ M $= \vec{I} = 0.5$ Ns, if M is the mass of the ball; thus, $M = 0.5/52 = 0.096$ kg $= 96$ g

(c) The work done by Natalie was $\bar{F}_{\text{horizontal}}x$, where x is the distance moved $= 0.9$ m. Thus, the work done is 500×0.9 m $= 450$ J.

What happens to Newton's 2nd law if the mass is not constant, as in a plane or in a rocket being launched, where the fuel is an appreciable part of the mass and is being exhausted? The mass also changes in relativity, as we shall see in a much later chapter. This is where the momentum becomes particularly useful. A more general expression than Newton's 2nd law is

$$\vec{F} = \frac{\Delta \vec{p}}{\Delta t} = \frac{\Delta(m\vec{v})}{\Delta t} = \bar{m}\frac{\Delta \vec{v}}{\Delta t} + \vec{\bar{v}}\frac{\Delta m}{\Delta t}, \tag{6.7}$$

where $\bar{m}$ is an average mass $= (m_i + m_f)/2$ and $\vec{\bar{v}}$ is an average velocity, $(v_i + v_f)/2$. We have used $\Delta(BY) = X\Delta Y + Y\Delta X$, and assume that ΔY and ΔX are much smaller than X and Y, respectively.

If the total or net force on an object vanishes, from Eq. (6.7), we see that its momentum is constant,

$$\Delta \vec{p} = 0. \tag{6.8}$$

This is another important conservation law. One of the reasons is that *internal* forces vanish because of Newton's third law. They vanish in pairs. The lack of change of momentum also agrees with Newton's first law.

When two objects are tied together, the force transmitted by the rope from one body to the other is an internal force of the system, composed of the two objects. The forces between molecules in a solid, liquid or gas are also internal forces. They cancel pairwise. It is only the total external forces which must vanish for the momentum to remain unchanged. If two people shove each other, this is also an internal force if we consider both people as composing the system; it is only if you consider each individual and the forces acting on each one, that the push is an external force. In a collision, it is generally possible to neglect external forces because the time involved is very short, so that momentum is conserved for the colliding system. In other words, to repeat, $\vec{p}_f = \vec{p}_i$ if $\vec{F}_{\text{ext}} = 0$. Note that it is *not* the velocity which remains invariant, but the momentum.

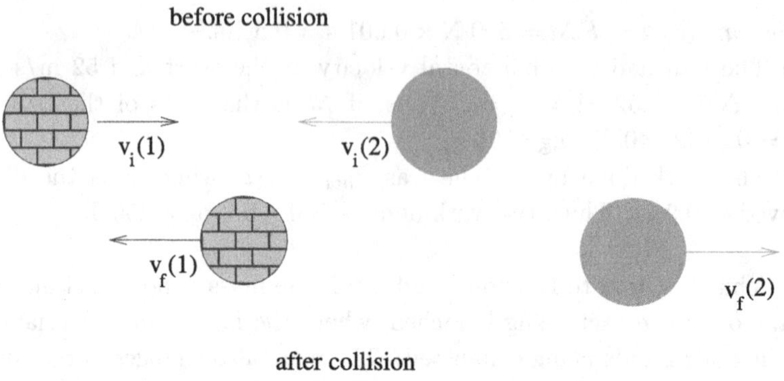

before collision

after collision

Fig. 6.2   A linear collision of two balls.

Like energy conservation, momentum conservation follows from a symmetry law. In this case it is the symmetry under spatial translation that is responsible. That is, an experiment can be done here or can be moved to a different location. If you translate an experiment by moving it, the momenta of the experiment remain unchanged. Momentum does not depend on position. Since the laws of physics do not depend on location, a translation leaves the laws of physics invariant. Momentum conservation can be shown to follow.

6.2.1   *Collisions*

First of all, we have to define "collision". It doesn't have to be violent, or sudden. It just has to be an encounter between objects, an interaction that has a beginning and an end, in a time interval short enough so that other interactions are relatively unimportant. The only important features are the two bodies and the forces they exert on each other. For example, if an asteroid comes from a great distance, sweeps by the Earth, and is gone in a day, that encounter can be treated as a collision between the asteroid and the Earth, because it occurs in a time that is short compared with the orbital periods of the Moon and the Earth.

The basic principle is: *in a collision, the total momentum is unchanged.* The reason is that external forces can be neglected during the collision.

As a specific example of momentum conservation, consider the collision of two bodies, such as two billiard balls or two cars. In both cases the external gravitational forces can be neglected during the time of collision.

It is only the internal forces that the two bodies exert on each other that matter. In that case, momentum is conserved for the colliding system and we have in an obvious notation

$$m_1 \vec{v}_i(1) + m_2 \vec{v}_i(2) = m_1 \vec{v}_f(1) + m_2 \vec{v}_f(2). \tag{6.9}$$

Illustrative Example: A bullet strikes a block of wood in which it gets embedded. If the bullet has a mass of 10 g and is originally moving at 200 m/s and the block has a mass of 1.0 kg and is initially at rest, what is the speed of the bullet and block after the collision?

Solution:

$$p_i = m_1 v_i(1) = 0.01 \text{ kg} \times 200 \text{ m/s} = 2 \text{ kg} - \text{m/s} = p_f = (m_1 + m_2)v_f$$

$$v_f = \frac{2 \text{ kg} - \text{m/s}}{1.01 \text{ kg}} = 1.98 \text{ m/s}. \tag{6.10}$$

Illustrative Problem: Consider truck A of mass 2.0×10^3 kg colliding with a stationary truck of 4.0×10^3 kg mass. Truck A was traveling at 1.25 m/s before the collision. What is the speed of the mess after the collision?

Solution: Momentum is conserved: $m_A v_i(A) = 2 \times 10^3$ kg $\times 1.25$ m/s $= (2+4) \times 10^3$ kg $\times v_f$.
Thus we have $v_f = 2.5/6 = 0.42$ m/s.

Illustrative Question: As a third example, consider the explosion of a ball at rest into 3 fragments of equal mass, as shown in Fig. 6.3. If fragment A flies off to the East and fragment B flies off to the South with the same speed, v, what is the direction and speed of the third fragment?

Answer: Since momentum is conserved, the third fragment must fly NW with a speed $= \sqrt{2}v$ so that the total momentum of the fragments adds to zero, the initial momentum.

Collisions are used by physicists to study atomic, nuclear, particle and solid object structures. Both energy and momentum conservation are used to help understand what happens in the collision and to get some insight into the structures of the colliding objects.

Momentum conservation is also related to Newton's third law. When two objects, A and B collide the $\Delta(m_A \vec{v}_A) = \vec{F}_{onA}\Delta t$ and $\Delta(m_B \vec{v}_B) =$

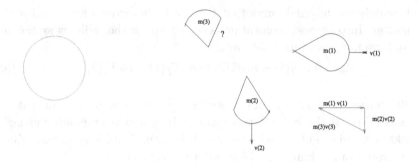

Fig. 6.3   Explosion of a ball into 3 fragments.

$\vec{F}_{onB}\Delta t$. Since momentum is conserved, we have $\Delta(m_A\vec{v}_A) = \vec{F}_{onA}\Delta t = -\Delta(m_B\vec{v}_B) = \vec{F}_{onB}\Delta t$, and therefore $\vec{F}_{onA} = -\vec{F}_{onB}$. The two quantities are equal and opposite so that there is no net force and no change in momentum.

If two particles are emitted by an object at rest, the two pieces must fly off in opposite directions with equal magnitudes of momenta. This is what happened in the beta decay mentioned in Chapter 5. When this was found not to be the case in a nuclear decay, physicists thought that a conservation law was broken. However, Wolfgang Pauli thought better of it and invented a third particle with no charge and very, very small (or no) mass to rescue the situation. He announced it at a conference to which he sent a letter because he was unable to attend. Why? He wanted to attend the annual ball at the University where he was employed! It took 30 more years to discover this elusive particle, called a neutrino (little neutral one).

It has been suggested that we can dodge a dangerous asteroid impact by moving the Earth out of the way, by means of captive rockets. A student problem will show that it is totally impractical. A more practical solution is to force the asteroid to change course.

6.2.2   *Balloons, squids, and rockets*

A rubber balloon is the simplest gadget for demonstrating the "rocket principle". Blow it up and let it go; the balloon collapses rapidly; it expels its compressed air, which drives the balloon in the opposite direction. A number of living creatures propel themselves by the rocket principle: the jellyfish, squid and octopus. The basic mechanism is: *In a closed system, the total momentum is constant.* A closed system is one with no external forces.

Quantitative relations are derived as follows. Assume a rocket of mass M, emitting a stream of gas at the rate of $\Delta m/\Delta t$ at a constant velocity $\mathbf{v}$ with respect to the rocket. Let the momentum of the rocket at time t be $\mathbf{P} = M\mathbf{V}$. In the interval Δt the rocket expels a quantity of exhaust gas Δm, which has momentum $\Delta m\mathbf{v}$ in the backward direction. The rocket experiences a change of momentum $\Delta\mathbf{P} = M\Delta\mathbf{V}$. Conservation of momentum dictates that $M\Delta\mathbf{V} + \Delta m\mathbf{v} = 0$. Dividing through by Δt, we get $M\mathbf{a} + (\Delta m/\Delta t)\mathbf{v} = 0$, where $\mathbf{a}$ is the rocket's acceleration. Thus, the exhaust gas exerts a driving force $\mathbf{F}_{\text{reaction}}$:

$$\mathbf{F}_{\text{reaction}} = -\mathbf{v}(\Delta m/\Delta t). \tag{6.11}$$

Illustrative Example: A 500 kg rocket is about to be launched for an interplanetary voyage. Its engines begin emitting a jet of hot gas straight downward, at a velocity of 100 m/s. What is the minimum exhaust rate (kg/sec) for liftoff?

Answer: The minimum exhaust rate must provide a force equal to the rocket's weight. Applying Eq. (6.11), $F_{\text{exhaust}} = 100 \times (\Delta m/\Delta t) = 500 \times 9.8$ N. Solving for the exhaust rate, we get $\Delta m/\Delta t = 49.0$ kg/s.

6.3 Center of Mass

When two objects collide, where is the *center of mass*? What happens to it during the collision?

We define the center of mass of two objects as the weighted average location of the two objects, where the weighting is according to their masses. In one dimension, if the objects are located at x_1 and x_2, we have

$$x_{\text{cm}} = \frac{m_1 x_1 + m_2 x_2}{m_1 + m_2}. \tag{6.12}$$

It should be clear how to generalize this definition to more than two objects or to 2 and 3 dimensions. We can also determine the velocity of the center of mass

$$\Delta x_{\text{cm}} = \frac{m_1 \Delta x_1 + m_2 \Delta x_2}{m_1 + m_2}, \tag{6.13}$$

$$\frac{\Delta x_{\text{cm}}}{\Delta t} = \frac{m_1 \Delta x_1/\Delta t + m_2 \Delta x_2/\Delta t}{m_1 + m_2}, \tag{6.14}$$

$$v_{\text{cm}} = \frac{m_1 v_1 + m_2 v_2}{m_1 + m_2} = \frac{p_1 + p_2}{m_1 + m_2}, \tag{6.15}$$

$$M_{\text{total}} v_{\text{cm}} = p_1 + p_2 = p_{\text{total}}, \tag{6.16}$$

where $M_{\text{total}} = m_1 + m_2$. If there is no net external force, then p_{total} remains constant and therefore v_{cm} does not change during a collision of two objects. If the center of mass is located at the origin at the start of a collision, it remains at zero during and after the collision.

Illustrative Question: In a faucet drip with many drops, is the center of mass of the drops halfway between the faucet and the sink, or is it higher or lower? Why?

Answer: The drops increase their speed as they fall. Thus there will be a larger distance between drops nearer the sink than nearer to the faucet and the center of mass will lie above the geometric middle.

In a later chapter we will study the center of mass of a solid extended object.

6.4   Energy and Collisions

We have seen that momentum is conserved in collisions as long as external forces can be neglected or cancel. Is kinetic energy also conserved? We know from experience that this is not the case, as when two cars collide. But kinetic energy may be conserved, as (approximately) in the collisions of two billiard balls. Collisions are grouped into three categories: *elastic collisions* are those for which the kinetic energy is conserved, *perfectly inelastic* collisions are those where the two colliding objects stick together after they collide. The only examples of perfectly elastic collisions are those between atomic or subatomic particles, when they suffer no change in their

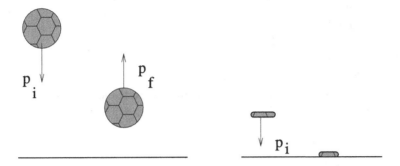

Fig. 6.4   An elastic collision and a totally inelastic one.

states (defined by quantum numbers). An example of a perfectly inelastic collision is the bullet and the piece of wood. The third category, simply called *inelastic collisions* or sometimes partially elastic, are those in which neither of the above conditions holds; for instance two cars may collide and bounce back. The totally inelastic collision leads to a maximum loss of kinetic energy. For an elastic collision, we have $K_f = K_i$; for an inelastic collision, $K_f < K_i$; for a totally inelastic collision $K_i - K_f$ is a maximum. In Fig. 6.4, we show an elastic collision, where $\vec{p}_f = -\vec{p}_i$ and a totally inelastic one, where $\vec{p}_f = 0$. The floor absorbs the difference of momenta, so that momentum is conserved in both cases.

Illustrative Problem: Two balls of mass 25 g collide head on and stick together. One ball was traveling at 2.0 m/s and the other one was at rest. Find the kinetic energy of the two balls after the collision and the ratio to the initial K.

Solution: $p_i = m(1)v_i = 25 \text{ g} \times 2 \text{ m/s} = 50 \text{ g} - \text{m/s} = p_f = (m(1) + m(2))v_f$.
50 g $\times v_f = 50$ g-m/s or $v_f = 1$ m/s.
The kinetic energy at the beginning was $K_i = \frac{1}{2}m(1)v_i^2 = 1.25 \times 10^{-2}$ kg $\times 4 \text{ m}^2/s^2 = 5 \times 10^{-2}$ J.
The final K was $K_f = \frac{1}{2}(m(1) + m(2))v_f^2 = 0.025$ kg $\times 1 \text{ m}^2/s^2 = 2.5 \times 10^{-2}$ J. Compare this to the initial K, $K_f/K_i = 2.5 \text{ J}/5 \text{ J} = 1/2$.

Is the total energy constant in a collision? Yes, even if the mechanical energy is not conserved, the total energy is always constant. When two objects collide and stick together, the kinetic energy goes into deformation energy, sound and heat.

Illustrative Problem: Consider two balls of equal masses on a table colliding elastically head on. If ball 1 has an initial speed v_i and ball 2 is at rest, show that after the collision ball 1 comes to rest and ball 2 moves off at velocity v_i.

Solution: Call the final velocities of the balls u_1 and u_2. Momentum and energy conservation give:

$$p_i = p_f, \tag{6.17}$$

$$mv_i = m(u_1 + u_2), \tag{6.18}$$

$$K_i = K_f, \tag{6.19}$$

$$\frac{1}{2}mv_i^2 = \frac{1}{2}m(u_1^2 + u_2^2). \tag{6.20}$$

Thus, we have

$$v_i = u_1 + u_2, \tag{6.21}$$

$$v_i^2 = u_1^2 + u_2^2. \tag{6.22}$$

Eliminate $u_1 = v_i - u_2$

$$v_i^2 = (v_i - u_2)^2 + u_2^2 = v_i^2 + 2u_2^2 - 2v_D u_2, \tag{6.23}$$

$$u_2^2 = v_i u_2, \tag{6.24}$$

$$u_2 = v_i. \tag{6.25}$$

There is another solution, namely $u_2 = 0$, but this is not sensible. Ball 2 does not stand still after it has been hit.

6.4.1   *Collisions in 2-dimensions*

So far we have discussed one dimensional collisions. The situation in 2-dimensions is a direct generalization of it. Momentum is a vector quantity and therefore momentum conservation must now be considered in 2-dimensions. If we consider the x-y plane as that of the collision, then we have

$$p_{ix}(1) + p_{ix}(2) = p_{PX}(1) + p_{PX}(2),$$

$$p_{Ly}(1) + p_{iy}(2) = p_{fy}(1) + p_{fy}(2),$$

$$\vec{p}_i(1) + \vec{p}_i(2) = \vec{p}_f(1) + \vec{p}_f(2). \tag{6.26}$$

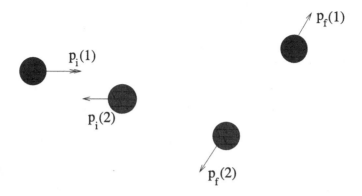

Fig. 6.5 A two-dimensional collision.

If the collision is elastic we also have

$$\frac{p_i^2(1)}{2m_1} + \frac{p_i^2(2)}{2m_2} = \frac{p_f^2(1)}{2m_1} + \frac{p_f^2(2)}{2m_2}. \tag{6.27}$$

You can use these three equations together to determine where one particle goes after the collision, if you know where the other one has gone or know the magnitude of its final momentum. The reason is that we have three equations and four unknowns $(p_{fx}(1), p_{fy}(1), p_{fx}(2), p_{fy}(2))$. These equations are often used in elastic collisions, and they can be modified for partially elastic collisions.

Illustrative Example: Consider two balls of masses $m_1 = 32$ gm and $m_2 = 48$ gm colliding at a glancing angle. They are traveling in opposite directions so that their total momentum is zero. Ball 1 goes off at an angle of 30° with a speed of 26.0 m/s. What is the velocity of ball 2?

Solution: The total momentum is zero so that the center-of-mass is at the origin and stays there. The momentum of ball 1 after the collision is 32 gm × 26 m/s = 0.832 kg-m/s. Ball 2 must appear at an gale of 210° with a speed of (0.832 kg-m/s)/0.048 kg = 17.3 ≈ 17 m/s, so that the total momentum remains zero after the collision.

Questions

Q6.1. When a gun is fired, should you hold the gun close to your shoulder or away from it? Why?

Q6.2. (a) You throw a ball against a wall and it rebounds with the same speed, but in the opposite direction. Is momentum conserved in the collision? Is mechanical energy conserved? Explain.
(b) Repeat part (a) with chewing gum which sticks to the wall.

Q6.3. For question 2, the masses for the two objects are the same. In which case, (a) or (b) is the impulse larger if the collision time is the same? Explain.

Q6.4. When two billiard balls collide, there is a small loss of kinetic energy. How can you tell that this is the case?

Q6.5. Which is safer to passengers in a car in a head-on crash: A car with a very strong front end or one with a crushable front end? Explain.

Q6.6. Can you think of a way of measuring the speed of a bullet by using collisions? How?

Q6.7. When a baseball bat strikes a baseball is momentum conserved? Explain.

Q6.8. Can a system of two objects have zero momentum but a finite kinetic energy? Explain. Is the reverse also possible? Explain.

Q6.9. In tennis as in golf, you are always advised to "follow through" on a swing. Why? Will the ball go further if you follow through? Explain.

Q6.10. A neutron (uncharged particle with almost the same mass as a proton or hydrogen atom) passes through a solid block of material and comes out in a different direction with a loss of energy. What can you conclude?

Q6.11. In a hydrogen gas, the molecules move in many different directions with an average speed of 480 m/s. What is the average momentum of the molecules?

Q6.12. You are on a train moving at 22 m/s and observe a collision of two objects on a platform. On the platform a ball is moving to the right at

20.0 m/s and collides with one standing still. How does the collision on the platform appear to you? Is momentum conserved, as seen by you? Explain your answers.

Q6.13. You run on a waxed floor, slide for a distance, and come to a stop. How can this be consistent with the law of momentum conservation?

Q6.14. You are trapped in the middle of a frozen lake. Friction is zero. How can you get off the ice? If you can think of a strategy, describe it quantitatively.

Q6.15. A rocket lifts off the surface of the earth directly upward. Does the earth recoil in the opposite direction? Explain.

Problems

P6.1. A ball of mass 10 g, moving with a speed of 12 m/s, collides with another one at rest with a mass 15 g. After the collision the 10 g ball moves backwards (in the direction it came from) with a speed of 2.3 m/s.
(a) Determine the velocity of the 15 g ball after the collision.
(b) Determine whether this is an elastic or inelastic collision.

P6.2. A child on a swing has a mass of 38 kg and accidentally hits an adult on the ground at the bottom of the swing's path. The adult has a mass of 76 kg. The child on the swing started her downward swing at a height of 3.0 m above the person's contact point, and comes to rest after hitting the adult on the ground.
(a) Determine the speed of the child just prior to hitting the adult.
(b) What is the speed with which the adult flies off?
(c) What is the kinetic energy of the adult?
(d) Is the collision elastic or inelastic? If inelastic, determine the amount of kinetic energy lost in the collision.

P6.3. A 66 kg man jumps off a bridge 12 m high just as a boat passes. He lands in the boat of mass 220 kg, which was moving at a steady rate of 3.0 m/s. What is the speed of the boat after the man lands in it?

P6.4. (Challenge) An alpha particle has a mass of 4 times that of the proton (1.67×10^{-27} kg). It collides elastically with another alpha particle at rest and flies off at 90° to its direction before the collision. Is it possible to conserve both momentum and kinetic energy in this collision? Explain.

P6.5. In the collision of Problem 4, the first alpha particle has a speed of 240 m/s before the collision and leaves the collision in a direction of 45° relative to its initial direction. Both particles leave the collision with equal speeds.
(a) What is the direction of the second alpha particle after the collision?
(b) Can the collision be an elastic one? Explain.
(c) Determine the velocities of both alpha particles after the collision.

P6.6. The two alpha particles of Problem 4 collide head on when both are traveling at the same speed of 240 m/s, but in opposite directions. One of them goes off at 90° to its original direction at 120 m/s.
(a) What is the velocity (magnitude and direction) of the second alpha particle?
(b) How much energy has been lost in the collision?

P6.7. A railroad car with a mass of 1200 kg collides and couples with a second car of mass 1800 kg that is initially at rest. The first car moves at 12 m/s due East prior to the collision.
(a) What is the initial momentum?
(b) What is the final velocity of the 2 cars after they couple?

P6.8. (Challenge) A marble of mass 22 g falls from a window shortly after a basketball of mass 540 g does so. Both have no initial velocity. The window is 22 m above the street. By the time the marble reaches the ball, the ball has just rebounded elastically from the street.
(a) Determine the speed with which the basketball hits the street.
(b) What is the speed of the marble just before it hits the basketball?
(c) The marble hits the basketball, which is moving upward, and rebounds from an elastic collision. If you neglect the distance the basketball has moved upwards and its diameter, will the marble rebound with a higher speed than it had just before the collision, smaller speed, or the same speed? Explain.
(d) Determine the rebound speed of the marble. (This involves the solution of a quadratic equation, but see the NOTE below).
(e) After the collision, will the marble go higher than, lower, or just 22 m? Find the height algebraically.
Note: You need only keep first order terms in the ratio of the marble to the basketball mass.

P6.9. A ball of mass 22 g rolls over the edge and bounces down a series of steps. Each step is at a height of 1/6 m above the previous ones. There are 12 steps to the floor below.
(a) What is the speed with which the ball reaches the floor below?
(b) How high will it rise on its bounce when it reaches that floor? Assume the collision is elastic.

P6.10. A bicyclist throws a ball of mass $m = 2.20$ kg forward at a speed of 3.20 m/s relative to the bicycle, while moving at a speed 12.0 m/s. The mass of the bicycle and individual is 124 kg. What is the speed of the cyclist immediately after the throw?

P6.11. A rocket moving off its launching pad has a mass of 7300 kg and attains a speed of 4800 m/s. Its lower part, of mass 3600 kg, separates and travels off at 5200 m/s towards the earth. What is the speed of the upper stage of the rocket after the separation?

P6.12. A large platform of mass 230 kg is floating in a lake. It has dimensions of 1.2 m by 4.5 m. A person of mass 68 kg swims to it, gets up on one end of it and runs from one end of the longer dimension to the other, reaching a speed of 3.6 m/s at the far end.
(a) With what speed will the platform move by the time the individual reaches the far end?
(b) A friend of mass 62 kg joins her at the far end and stays put while the first person runs back to the other end again. Will the platform return to its original position again? If not, at what speed will it move this time?

P6.13. In the demonstration of eight billiard balls in a row, it was shown that if one ball collides with the seven others, then one flies off, but two fly off if two balls are used to collide with the other six. Show that you cannot have two billiard balls fly off in an elastic collision if only one ball is incident.

P6.14. Two cars collide at an intersection. The first car, of mass 1520 kg was heading straight South at 65 mph, whereas the second one, of mass 2210 kg was headed straight West at a speed of 42 mph. The two cars stick together after the collision.
(a) Find the velocity with which the cars leave the collision.
(b) Find the initial and final kinetic energies and determine K_f/K_i.

P6.15. A 1500 kg car traveling due N with a speed of 25 m/s collides head on with a 4500 kg truck traveling due S with a speed of 15 m/s. The car rebounds backwards at a speed of 10 m/s and the truck keeps going forward.

(a) Find the speed of the truck after the collision.

(b) Is the collision elastic, inelastic, or perfectly inelastic? Give reason.

P6.16. You (mass m) are in a space ship (mass M), traveling at velocity $\mathbf{V}$ with respect to Earth. You walk, at velocity $\mathbf{v}$ relative to the ship, from one end of the ship to the other. What is the velocity of the ship, as observed from Earth, while you are walking?

P6.17. An artillery shell of mass M is moving at velocity $\mathbf{v}$ bursts into three fragments while it is in the air. What is the total momentum of the three fragments immediately after the burst?

P6.18. A ball carrier (mass M), running at speed v, is tackled by a guard (mass $1.5\ M$) traveling in the opposite direction at speed $0.52\ v$. After the collision, while the two are coupled together, how fast do they travel and in what direction?

P6.19. You are running in the hallway at speed v, late for class, when you collide with a teacher, whose mass is 3 times yours, walking in the same direction, at speed $0.30\ v$. You bounce backward, at speed $0.50\ v$. Describe the teacher's motion immediately after the collision.

6.20. A railroad freight car is taking on a load of coal while it travels at 5 km/hr, through the loading facility. The car is loading at the rate of 2000 kg/minute. How much force is needed to keep the car moving steadily?

P6.21. A dump truck is leaking sand at the rate of 0.10 kg/s, while it is traveling at 50.0 km/hr. How much force is exerted by the leaking sand?

P6.22. You are running in a rainstorm. The rain is coming straight down, you are running at speed v, impacting rain at the rate of $\Delta m/\Delta t = 0.012$ kg/s, and the rain drains off you as you run.

(a) Does the rain exert a force on you?

(b) If yes, how large a force?

Answers to odd-numbered questions

Q1. Close. The mass then includes that of the person as well as the gun.

Q3. (a) Because the momentum change is larger.

Q5. Crushable front end, so that the time of collision is longer.

Q7. Yes, for the combination of ball, bat, and batter.

Q9. The time of impact is longer, so that the change of momentum is larger.

Q11. 0.

Q13. You need to consider the entire system, not just one part of it.

Q15. Yes, since momentum of the earth + rocket is conserved.

Answers to odd-numbered problems

P1. (a) From momentum conservation, $v = 6.5$ m/s.
(b) $E_f < E_i$; thus the collision is inelastic.

P3. From momentum conservation, $v = 2.3$ m/s.

P5. (a) Also 45°, but SE is the first one was NE.
(b) Yes.
(c) $v = 170$ m/s.

P7. (a) 1.44×10^4 kg-m/s.
(b) $v = 4.6$ m/s due E.

P9. (a) 6.3 m/s.
(b) 2 m.

P11. From momentum conservation, $v = 14.5$ km/s.

P13. It is not possible to conserve both momentum and kinetic energy.

P15. (a) 3.3 m/s.
(b) Inelastic $(K_f < K_i)$.

P17. M**v**.

P19. $\vec{p}_i v = 1.9 m\vec{v} = \vec{p}_f = 3m\vec{v}_{\text{teacher}} - 0.5m\vec{v}$; thus $\vec{v}_{\text{teacher}} = 0.8\vec{v}$.

P21. $F = \Delta p/\Delta t = (\Delta m/\Delta t)v = (0.1 \text{ kg/s})(50/3600 \text{ km/s}) = 0.0014$ N.

Chapter 7

Rotational Dynamics

Aristotle taught that rotational motion is divine, because it could continue forever, whereas linear motion must have a beginning and an end. We won't go so far as to call it divine, but rotational motion is very important to us.

Many familiar objects go around and around in circles about an *axis of rotation*. Examples are car and bicycle wheels, windmills, water-wheels, washing and drying machines, records and DVDs, and many more examples.

7.1 Circular Motion, Centripetal Acceleration

When a particle moves in a circle, its direction keeps changing; thus it is *accelerating*, even if it moves at constant speed. The direction of the acceleration can be found by considering two neighboring points on the circle.

A convenient unit for angle measurement is the *radian*. The measure of an angle in radians is defined in terms of a circular arc, the ratio of arc length to the radius of the circle: see Fig. 7.1. Note that the radian measure has no units; it is a pure number.

$$angle\ in\ radians\ \theta = d/R; \quad d = R\theta\,,$$

where θ is expressed in radians. A complete circle has a circumference $2\pi R$, therefore its radian measure is 2π. In terms of degrees, a full circle corresponds to a rotation of $360°$; therefore

$$1\ radian = 360°/2\pi = 57.3°\,.$$

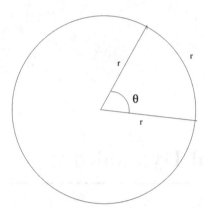

Fig. 7.1   Picture of 1 radian.

Rotational speeds are given in a variety of units, e.g. *rad/sec*, *rev/sec*, *rev/min*; often for motor driven machinery, *rpm*, for revolutions per minute. We will use *rad/sec*.

Illustrative Example: The shaft of an electric motor is rotating at 3600 rpm. Calculate its speed in radians/sec.

Solution: 3600 rpm × 1/60 min/sec × 2π radians/revolution = 377 rad/sec.

When only the direction of motion changes, it involves an acceleration perpendicular to the path, even when the speed is constant along the path. Figure 7.2 shows a particle moving at constant speed along a curved path, so that the velocity vector changes from $\mathbf{v}$ to $\mathbf{v}'$ as the particle advances a short distance d. As the diagram shows,

$$\mathbf{v}' = \mathbf{v} + \Delta\mathbf{v}.$$

The curved path can be fitted locally to a circle whose radius is R. Then R and d define an angle

$$\Delta\theta = d/R.$$

The magnitude of the change in velocity can be described in terms of this angle:

$$\Delta v = v\Delta\theta = v(d/R).$$

Since $d = Pt$, we have

$$\Delta v = v(OT/R).$$

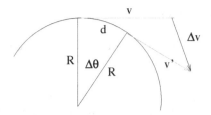

Fig. 7.2 Centripetal acceleration for a particle moving along an arc at constant speed.

Dividing through by t, we get

$$\Delta v/t = v^2/R.$$

In the diagram $\Delta \mathbf{v}$ is not quite perpendicular to $\mathbf{v}$, but it approaches a right angle as the interval decreases. Thus, the change of direction produces an *acceleration that is locally perpendicular to the path, along the radius of curvature, toward the center of the circle,* as pictured in Fig. 7.2. It is called a *centripetal acceleration* because centripetal means seeking the center.

The equation for centripetal acceleration is:

$$a_c = \frac{v^2}{R}.$$

To emphasize, the centripetal acceleration is due to the change in direction, not a change in speed.

Illustrative Example: Calculate the centripetal acceleration of a point on the rim of a 27 inch diameter bicycle tire rotating at 30 rpm.

Solution: Convert the angular velocity:

$$30 \text{ rpm} \times (1/60) \times 2\pi = 3.14 \text{ rad/sec} = \omega.$$

$$a_c = (3.14)^2 \times 13.5 \times 2.54 \text{ cm/inch} = 338 \text{ cm/s}^2 \approx 340 \text{ cm/s}^2.$$

A force is required to keep a particle moving in a circle, even at constant speed, since there is an acceleration from the change of direction of the motion.

7.2   Rotational Motion of Solid Objects

How do we describe rotational motion of a rigid extended object? We need to define a *rotational displacement*, a *rotational velocity*, and a *rotational acceleration*.

We describe rotational velocity akin to the linear one. Linear velocity $\vec{v} = \vec{d}/t$, where $\vec{d}$ is the linear displacement. Rotational or *angular speed*, ω is defined similarly for an *angular displacement* θ

$$\omega = \frac{\theta}{t}, \tag{7.1}$$

where θ is in radians, so that the angular speed is given in rad/s. The linear velocity is a vector. Is there a correspondence for circular motion? Yes, the *angular velocity* has a direction given by the right-hand rule: The angular velocity is along the direction of the thumb of the right hand (along the axis of rotation) if the fingers curl in the direction of motion. For a counter-clockwise rotation in a horizontal plane, the angular velocity is upward along the axis. For a clockwise rotation, it is downward. The length of an arc is θr, if θ is given in radians. We thus obtain a relation between linear and angular speeds; whereas the linear speed $v = d/t$, the *angular* speed is $\omega = \frac{\theta}{t}$:

$$d = r\theta,$$

$$\frac{d}{t} = r\frac{\theta}{t},$$

$$v = r\omega$$

$$a_c = v^2/r = \omega^2 r \tag{7.2}$$

In analogy to the linear case, we define angular acceleration, α, as the rate of change of angular velocity per unit time,

$$\vec{\alpha} = \frac{\Delta\vec{\omega}}{t} \tag{7.3}$$

where $\Delta\omega = \omega_f - \omega_i$ is the change of angular velocity. α is measured in rad/s^2. The direction of α is along the axis of rotation, parallel to $\vec{\omega}$ if the angular velocity is increasing and anti-parallel if it is decreasing. The linear correspondence to α is the tangential acceleration, $a_T = \alpha r$ is not along the axis, but in the direction of the velocity if $a_T \geq 0$ and opposite to it if $\alpha \leq 0$, as shown in Fig. 7.3. Note that $\vec{a}_T$ is not parallel to α; the latter is along the axis, whereas a_T is tangent to the rotation. It is important to note the difference between the tangential acceleration and

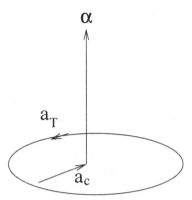

Fig. 7.3 a_c, α and a_T for a particle moving clockwise at increasing speed.

the centripetal acceleration. The latter occurs even at a constant rotational speed, whereas the tangential acceleration requires a change of that speed. Both accelerations can exist at the same time and they are at right angles to each other (see Fig. 7.3), so that the magnitude of the total acceleration is

$$a = \sqrt{a_T^2 + a_c^2}\,. \tag{7.4}$$

If $a_T \neq 0$, then $a_c = v^2/r = \omega^2 r$ keeps changing with time.

To summarize, we have the correspondence between linear and circular motions:

$$d \to \theta$$

$$v \to \omega$$

$$a \to \alpha_T$$

$$angular\ displacement = \theta = d/r\,,$$

$$angular\ velocity = \omega = \theta/t = v/r\,,$$

$$angular\ acceleration = \alpha_T = \Delta\omega/t = a_T/r\,. \tag{7.5}$$

These relations make it clear that the displacement, the linear speed, and the linear acceleration all increase linearly with the distance from the rotational axis.

We have studied linear motion earlier and we saw a number of relations which can be taken over, with minor modifications, for circular motion. For a constant linear or rotational acceleration, we have:

$$v = v_i + at \qquad\qquad \omega = \omega_i + \alpha_T t \qquad\qquad (7.6)$$

$$d = v_i t + \frac{1}{2}at^2 \qquad\qquad \theta = \omega_i t + \frac{1}{2}\alpha_T t^2 \qquad\qquad (7.7)$$

Illustrative Problem 7.1: A small marble, with mass = 25 g, accelerates as it spins in a sink with sloping sides. When the diameter of the sink is 50 cm, its rotational speed is 2 rad/s and its tangential acceleration is 1.2 rad/s^2. (a) Find the tangential and the magnitude of the total acceleration in m/s^2. (b) Find the angular displacement during the next second.

Solution: (a) The tangential acceleration $a_T = \alpha r = 1.2$ rad/s^2 × 0.25 m = 0.3 m/s^2. The total acceleration is

$$a = \sqrt{a_T^2 + a_c^2}\,;$$

$$a_c = v^2/r = \omega^2 r = (2 \text{ rad/s})^2 \times 0.25 \text{ m} = 1 \text{ m/s}^2\,;$$

$$a = \sqrt{1 + 0.09} = 1.04 \text{ m/s}^2.$$

(b) $\qquad \theta = \dfrac{1}{2}\alpha t^2 + \omega_0 t = \dfrac{1}{2}1.2 \times 1^2 \text{ rad} + 2 \text{ rad/s} \times 1 \text{ s} = 2.6 \text{ rad.}$

7.3   Vector Multiplications

We need to digress in order to define the equivalent of a force for rotational motion.

There are several ways to multiply two vectors together. We shall be concerned with two of these. The first one is called *scalar multiplication* or *dot product*. The dot product of two vectors $\vec{A}$ and $\vec{B}$ is a scalar C,

$$C = \vec{A} \cdot \vec{B} = AB \cos\theta, \qquad\qquad (7.8)$$

where θ is the angle between $\vec{A}$ and $\vec{B}$ (see Fig. 7.4(a)). We have already made use of this multiplication, without naming it, when we evaluated the work done by a force $\vec{F}$ in displacing an object through distance $\vec{d}$,

$$W = \vec{F} \cdot \vec{d} = Md \cos\theta. \qquad\qquad (7.9)$$

There is another way to multiply two vectors together; it gives rise to a new vector, $\vec{V}$ and is called *vector multiplication* or *cross product*. For the two vectors $\vec{A}, \vec{B}$, we obtain

$$\vec{V} = \vec{A} \times \vec{B} = -\vec{B} \times \vec{A}. \qquad\qquad (7.10)$$

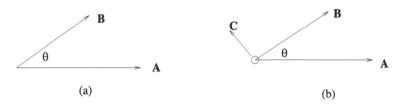

Fig. 7.4 The dot (a) and cross (b) products of **A** and **B**. The direction of C is perpendicular to the paper; the small circle means upwards.

The magnitude of $\vec{V}$ is given by $AB\sin\theta$, where θ again is the angle between the vectors $\vec{A}$ and $\vec{B}$. The direction of $\vec{V}$ is perpendicular to both $\vec{A}$ and $\vec{B}$; it is in the direction a right hand screw would advance if $\vec{A}$ is turned towards $\vec{B}$; see Fig. 7.4(b). Alternatively, you can curl your right hand from $\vec{A}$ towards $\vec{B}$; your thumb is then in the direction of $\vec{V}$. We shall see an application in the next section.

7.4 Torque

A force is required to produce a linear acceleration. What is it that produces a rotational acceleration? It is not enough to know the force; we also need its point of application relative to the axis of rotation. We know that it requires more of a push to open a door if we apply the push nearer to one of the hinges than closer to the edge of the door. It is a *torque* that is required. The torque is the force multiplied by the *lever arm*, where the lever arm is the perpendicular distance from the axis of rotation to the line of action of the force,

$$|\vec{\tau}| = |\vec{F}|r_\perp, \qquad (7.11)$$

where the Greek letter "tau", τ, is the conventional symbol for torque, $r_\perp$ is the lever arm; see Fig. 7.5. We use $|\vec{F}|$ in Eq. (7.23) because the torque is not in the direction of $\vec{F}$. The lever arm, $\vec{r}_\perp$ is defined in Fig. 7.5. It is the distance from the axis of rotation to the force along a line that is perpendicular to the applied force. If the force is applied perpendicular to the radius, that is along a tangent, the lever arm is simply the radius, as in Fig. 7.5(a). If it is applied at some other angle, as in Fig. 7.5(b), then it is $r\sin\theta$.

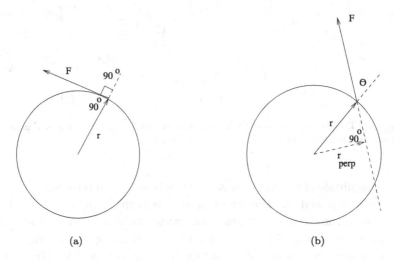

Fig. 7.5   Illustrations of the moment arm; (a) $r_\perp = r$, (b) $r_\perp < r$. The direction of the vectors $\vec{r}, \vec{r}_\perp$ is from the axis of rotation outward.

With the introduction of cross products, we can now define $\vec{\tau}$ more concisely

$$\vec{\tau} = \vec{r} \times \vec{F}, \tag{7.12}$$
$$|\vec{\tau}| = rF \sin\theta,$$

To come back to the door, if we push along the plane of the door to open it, we will never succeed; $\vec{\tau} = 0$ because $\sin\theta = 0$. If we push perpendicular to the door, we know that this works best; here $\theta = 90°$ and $\sin\theta = 1$. A wrench, when used to tighten a nut is an example of a simple machine with the lever arm equal to the length of the wrench from the nut to where the hand is located. The force is applied at right angle to the length of the wrench. The longer the wrench, the smaller the force that is required to turn the nut with a given torque.

Illustrative Problem 7.2: A force of 3 N is applied to a small object of 30 g at an angle of 60° to the radius vector of 10 cm, as shown in Fig. 7.5(b). Find the magnitude and direction of the torque due to this force.

Solution: The direction of the torque follows by turning the radius vector ($\vec{r}$) towards the force. A right handed screw would then advance out of the page, and this is the direction of the torque. Its magnitude is $rF \sin\theta = 0.1$ m $\times 3$ N $\times \sin 60° = 0.3 \times 0.867 = 0.26$ Nm.

7.5   Moment of Inertia

For linear motion we know that the acceleration is $\vec{a} = \vec{F}/m$, where m is a measure of the inertia of the body. What is the equivalent quantity for rotational motion? It is somewhat more complex. The *rotational inertia* or *moment of inertia*, as it is often called, depends on the distribution of mass of the body relative to the location of the axis of rotation. The closer to the axis that most of the mass is located, the smaller the inertia to a change of angular velocity.

Consider, first, a small mass, m, rotating about an axis at a distance r. From Newton's 2nd law, we have $\vec{F} = m\vec{a}$. We can rewrite this in terms of the torque as

$$|\vec{\tau}| = |\vec{r} \times \vec{F}| = mra = (mr^2)(a/r) = mr^2\alpha, \qquad (7.13)$$

$$\vec{\tau} = I\vec{\alpha} = mr^2\vec{\alpha}, \qquad (7.14)$$

$$\vec{\alpha} = \vec{\tau}/I. \qquad (7.15)$$

The quantity I is called the *moment of inertia* or rotational inertia of the body and for rotational motion replaces the mass of linear motion. In the above example $I = mr^2$.

In order to find the moment of inertia of an extended object we need to subdivide the object into tiny pieces and add the moments of inertia of all the small pieces,

$$I = \sum_i m_i r_i^2, \qquad (7.16)$$

where m_i is the mass of each small piece and r_i is its distance from the axis of rotation. We can think of each piece as an atom. For instance, for a linear rod rotating about one end, we can subdivide the length into tiny segments and add all the moments of inertia of the pieces about this axis.

Some examples of the moment of inertia for regular bodies are given in Fig. 7.6. It is less difficult to rotate a rod about its center than through one of its ends because the moment of inertia is smaller for a rotation about its center, as shown in Fig. 7.6; there is less mass at larger distances from the axis than when the rod rotates about one of its ends. The rotational inertia is $\frac{1}{3}mL^2$ for a rotation about one end, but it is 4 times smaller for a rotation about the center, namely $\frac{1}{12}mL^2$; here m is the mass of the rod, assumed to be uniform, and L is its length. The more mass is concentrated further away from the axis of rotation, the larger the moment of inertia.

If the axis of rotation is not at the center of symmetry of a body of uniform density, but is displaced by a distance ℓ from it, the moment of

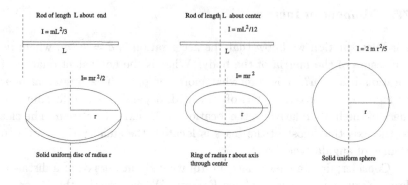

Fig. 7.6   Moments of inertia of various regular bodies about the dashed lines.

inertia is increased by $m\ell^2$,

$$I = I_0 + m\ell^2 \,, \tag{7.17}$$

where I_0 is the moment of inertia about the center of mass. For example, the moment of inertia of a linear rod of uniform density is $\frac{1}{12}mL^2$ about the center. It is $\frac{1}{3}mL^2$ about one end. The displacement of the end from the center is $\ell = L/2$ and the moment of inertia about one end is thus $\frac{1}{12}mL^2 + \frac{1}{4}mL^2 = \frac{1}{3}mL^2$.

For a bicycle wheel with a uniform thin rim, if you can neglect the mass of the spokes, the moment of inertia is simply mr^2, whereas for a solid wheel of the same radius it is only $\frac{1}{2}mr^2$, since more mass is located closer to the center.

Sometimes, it is easy to change the moment of inertia. A twirling ice skater does just that in pulling in her arms. She decreases her moment of inertia. For a constant torque this increases the rotational acceleration.

7.6   Angular Momentum

For rotational motion, what is the analogue to linear momentum? It is *angular momentum*. Sometimes *angular momentum* is referred to as *rotational momentum*. The linear momentum is $m\vec{v}$. We therefore expect the angular momentum, $\vec{L}$, to be

$$\vec{L} = I\vec{\omega} \,. \tag{7.18}$$

This is correct. For a small mass the angular momentum can be related to the linear momentum,

$$\vec{L} = \vec{r} \times \vec{p}, \tag{7.19}$$

where r is the distance of the object from the axis of rotation. We can see the connection of Eq. (7.19) to Eq. (7.18) for the circular motion of a small body. Namely, $I\omega = (mr^2)(v/r) = mvr = pr$.

The direction of the angular momentum is given by that of $\vec{\omega}$. This follows from Eq. (7.18).

Illustrative Problem: If, in Illustrative Problem 7.2, the mass is a small sphere of radius 2 cm (the center of the sphere is on the circumference of the circle of radius 10 cm), find the angular acceleration of the rotation about the center of the circle.

Solution: The moment of inertia of a sphere about its center is $I_{cm} = \frac{2}{5}mr^2 = \frac{2}{5}0.03$ kg$(0.02)^2$ m$^2 = 4.8 \times 10^{-6}$ kgm^2. Its moment of inertia about the center of the circle is then $I = mR^2 + I_{cm}$, where R is the radius of the circle. Thus $I = 0.03$ kg $\times (0.1)^2 + 4.8 \times 10^{-6} = 3 \times 10^{-4}$ kgm^2 + $4.8 \times 10^{-6} = 3.05 \times 10^{-4}$ kg/m^2 and the angular acceleration is $\alpha = \tau/I = 0.26$ Nm$/(3.05 \times 10^{-4})$ kgm$^2 = 8.5 \times 10^2$ rad/s^2.

Illustrative Problem: In Illustrative Problem 7.1, find the angular momentum of the marble when it is at a diameter of 50 cm.

Solution: The moment of inertia of the marble about the center of the circle is $I = mr^2 = 0.025$ kg$(0.25$ m$)^2 = 1.56 \times 10^{-3}$ kgm^2. The angular momentum's magnitude is $L = I\omega = 1.56 \times 10^{-3}$ kgm$^2 \times 2$ rad/s $= 3.12 \times 10^{-3}$ kgm^2/s.

If I is not constant, Eq. (7.14) cannot be used. The generalization is similar to that for linear motion, where $\vec{F} = \frac{\Delta \vec{p}}{t}$. This equation replaces $\vec{F} = m\vec{a}$.

We have, then,

$$\vec{\tau} = \frac{\Delta \vec{L}}{t} = \frac{\Delta(I\vec{\omega})}{t}. \tag{7.20}$$

For a tiny object, or for a wheel with all its mass concentrated on the rim, and rotating about an axis through the center and perpendicular to the plane of the wheel, this can readily be shown to be true.

For linear motion, linear momentum is constant (conserved) if there is no external force. If there is no external torque, it follows from Eq. (7.20) that the angular momentum is constant. In other words, angular momentum is constant if there is no net external torque. If $\vec{\tau} = 0$, and the moment of inertia changes, the angular velocity must also be altered to compensate for this change. An ice skater makes use of this when she goes into a spin, as does a diver when she goes into a tuck. The constancy of angular momentum can also be demonstrated by sitting a person on a rotating stool with some weights in her hands. If the person is set in rotation at constant speed with her arms tucked in, and then extends her arms, she will slow down considerably. If she pulls her arms back in again, she will speed up to the original speed, if friction is neglected.

Angular momentum conservation also follows from a symmetry principle. Because space is uniform, we can do an experiment and repeat it in a laboratory that is rotated from the original one. The same results will be obtained. This rotational independence gives rise to angular momentum conservation in the absence of a torque.

Note that the angular acceleration is in the same direction as $\vec{\omega}$ when the rotational speed is increased, but in the opposite direction when $\vec{\omega}$ is decreased. As a summary, we note the analogies to linear motion

$$m \to I$$
$$\vec{F} \to \vec{\tau}.$$
$$\vec{p} \to \vec{L} \qquad\qquad (7.21)$$

The angular momentum of a moving bicycle ($\vec{L}$ in Fig. 7.7) is perpendicular to its motion, in a horizontal direction, as long as the bicycle is upright. If it starts to fall, say to the right, then the angular momentum acquires a component (L'' in Fig. 7.7) in the forward direction, since the axis of rotation for the falling motion is forward and backwards. The total angular momentum then has a component horizontally perpendicular to the motion and also acquires a component forwards. If the cyclist now turns to the right, she will compensate for the change in angular momentum and the bicycle will right itself again. The faster the bicycle is going, the less of a turn it requires. It is hard to change the angular momentum direction of a rapidly rotating wheel.

This fact can be demonstrated in class by sitting a student on a stool and handing her a rotating bicycle wheel with a handle protruding from the axis. If the stool the student is sitting on is free to rotate, then when the student tries to turn the wheel upside down, the stool and student will rotate in

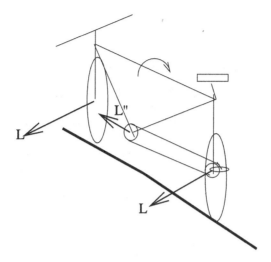

Fig. 7.7   A falling bicycle.

the opposite direction in order to keep the angular momentum constant. Although there is a torque on the wheel there is no external torque on the system of student-stool-wheel, so that the angular momentum of the system has to remain constant. The student will also experience the difficulty of changing the angular momentum of the wheel.

Many of you have probably played with a top. It works on the same principle. As long as the top is spinning fast, it is difficult to change its angular momentum. As it slows down and begins to fall, the direction of the angular momentum changes. This causes the axis of rotation of the top to change in order to compensate, and leads to a precession of the top.

7.7   Rotational Kinetic Energy

The kinetic energy of a moving body is $K = \frac{1}{2}mv^2$. We therefore expect the analog for rotational motion to be $K = \frac{1}{2}I\omega^2$, and this is again correct. As for linear motion, where the change in kinetic energy is equal to the work done, this is also the case here. For linear motion, the work is just $F_\parallel d$, where $F_\parallel$ is the component of the force parallel to the motion and d is the distance moved. For rotational motion, the work done is then expected to be

$$W = \tau\theta. \tag{7.22}$$

This can be shown to be correct:

$$W = F_{\|}d = (r_{\perp}F_{\|})(d/r_{\perp}) = \tau\theta \qquad (7.23)$$

The work is also equal to the change of kinetic energy,

$$\tau = I\alpha, \qquad\qquad \theta = \Delta\omega^2/(2\alpha), \qquad (7.24)$$

$$W = \tau\theta = (I\alpha)\left(\frac{\Delta\omega^2}{2\alpha}\right) = I\Delta\omega^2/2 = \Delta K. \qquad (7.25)$$

We have assumed that $\theta = 0$ at the start. Thus, for rotational motion we have

$$K = \frac{1}{2}I\omega^2. \qquad (7.26)$$

We summarize the analogies:

$$W = F_{\|}d \to W = \tau\theta \qquad (7.27)$$

$$K = \frac{1}{2}mv^2 \to K = \frac{1}{2}I\omega^2 \qquad (7.28)$$

$$K = \frac{p^2}{2m} \to K = \frac{L^2}{2I} \qquad (7.29)$$

Note the last entry.

Illustrative Problem: In Illustrative Problem 7.1, what is the kinetic energy of the marble at a diameter of 50 cm?

Solution: $K = \frac{1}{2}I\omega^2 = \frac{1}{2}L\omega = 0.5 \times 1.25 \times 10^{-2}$ kgm^2/s $\times$ 2 rad/s $= 1.25 \times 10^{-2} J$.

7.8   Center of Gravity

We have discussed the center of mass in a collision of two bodies in Chapter 6. Here we are dealing with forces and torques and therefore discuss the *center of gravity* of an object, which is the point where its weight can be considered to be located so that no net torque is exerted due to its weight. It is at the same location as the center of mass.

For an object of uniform density, the center of gravity is located at its geometric center. An example is a disc, a plank, or a rod, all of uniform density. The center-of-gravity need not be located in the object. For instance, a disc with a hole in the center, has the center of gravity located at the empty center. You can locate the center of gravity of an object by trying to balance it at a point. For an irregular planar object, you can also locate the center of gravity by suspending the object from at least two locations on the object and finding the intersection of two vertical lines,

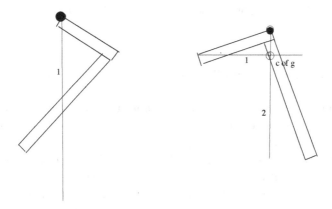

Fig. 7.8   Location of the center of gravity by suspension.

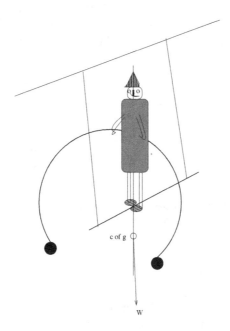

Fig. 7.9   A toy clown on a tight-wire (note the location of the center of gravity).

each one starting at the suspension point. The center of gravity must lie below both points of suspension, so that the moment arm, $r_\perp$ is zero. See Fig. 7.6. Galileo was first recognized for demonstrating how to locate the center of gravity of an irregular object.

Some toys work on the stability of an object when the center of gravity lies below the center of rotation. An example is the clown on a tight-wire, Fig. 7.9.

7.9   Equilibrium Conditions

If an object can rotate, then, for equilibrium it is not enough for the net force on it to be zero, but it is also necessary that there be no net torque. Take a meter stick as an example. To support it at a point, we need the support placed at the center, if its density is uniform. In that case, the right half of the stick leads to a torque in the clockwise direction (the direction is into the paper in Fig. 7.10), but the left half of the stick has an equal, but oppositely directed torque in the counterclockwise direction, or out of the paper. If you hang a 100 g mass at the 75 cm mark, you can balance it by a 100 g mass at 25 cm or a 50 g mass at 0 cm (see Fig. 7.10). In both cases, the net torque about the support is zero. This assumes that the suspension will bear and counter the added weight, so that the net force also vanishes.

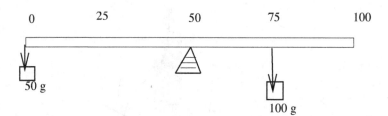

Fig. 7.10   Equilibrium of a meter stick with two masses. The arrows indicate the direction of the weights corresponding to the masses.

7.10   Rolling Motion

We consider rolling without sliding. Examples are bicycle wheels and automobile tires on a road with sufficient friction. In these cases, there is both rotational and linear motion. The two are related as can be seen from the

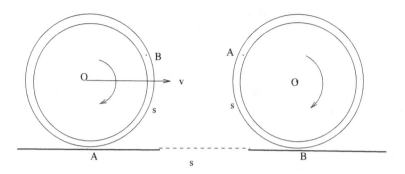

Fig. 7.11   Rolling without slipping.

figure of a bicycle wheel which advances while it rotates. The relationship is shown in Fig. 7.11, where it is seen that when the wheel advances a distance corresponding to the arc length AB it has turned through an angle $\theta = \frac{arc\ length}{r} = \frac{s}{r}$;

$$v = \omega r, \tag{7.30}$$

where v is the linear velocity of the center of the wheel and r is the wheel's radius.

The kinetic energy of a rolling wheel is made up of two parts, that due to translation and that due to rotation:

$$K = \frac{1}{2}mv^2 + \frac{1}{2}I\omega^2, \tag{7.31}$$

where I is taken about the center of mass. The point of contact of the wheel and road is actually stationary instantaneously. Thus, it is static friction which comes in here. (See, e.g. Fig. 7.11).

Questions

Q7.1. Can an extended object have a net torque if the net force is zero? Can the object have a net force if the torque is zero? Explain.

Q7.2. Is the angular momentum of an object always conserved? Explain.

Q7.3. Two go-carts are rolling down hill. Does the one with larger wheels have an advantage? The mass of the cart is small compared to that of the wheels. Explain your answer.

Q7.4. Two crates are full of junk. One has the heavier stuff nearer to the top than the other one. Which one has a larger moment of inertia for tipping it over? Explain.

Q7.5. A child on a merry-go-round moves out from the center towards the edge. Will the merry-go-round slow down, speed up, or continue at the same angular speed? Explain.

Q7.6. In Question 5, what is the net force? What is the moment arm? Explain.

Q7.7. The earth rotates about its own axis. Where, on the surface of the earth, is the linear velocity a minimum?

Q7.8. Even when a car is traveling at a constant velocity, the rims of the four tires are accelerating. Explain.

Q7.9. The speedometer in a car is calibrated in miles/hour, but it is based on the angular velocity of the wheels. If you change tires to slightly larger ones, and the car travels at the same speed, will the speedometer read the same, less, or more than before? Explain.

Q7.10. You are balancing on a stretched rope. Is there an advantage to bending down when doing so? If so, why? If not, why not?

Q7.11. If a person falls off a merry-go-round, what happens to its speed? Why?

Q7.12. Explain how a yo-yo works. Where does the torque for its acceleration come from?

Q7.13. Two cylinders roll down the hill. The cylinders have the same radii and the same masses, but one is hollow and the other one is solid. Which one reaches the bottom first? Why?

Q7.14. Two cylinders of the same radius and mass are on an incline. One rolls down without sliding and the other one slides down without rolling. Which one reaches the bottom first? Why?

Problems

P7.1. A pulsar is a neutron star that may rotate about a twin or another star and sends radio signals, which can be intercepted on earth. The radio signals are like the beam of a lighthouse that faces the earth once each revolution. If the time between signals is regular and is 0.35 s, what is the angular speed of the pulsar?

P7.2. For a small object moving in a horizontal circle, show the correctness of Eq. (7.20); derive $\tau = I\Delta\omega/t$.

P7.3. Light travels at 3×10^8 m/s. A laser light goes through a notch in a rotating toothed wheel. The light is reflected by a mirror 20 km away and returns. How fast must the wheel turn for the reflected light to pass through the next notch, if the notches are spaced $4°$ apart?

P7.4. (Challenge Problem) Consider a solid cylinder rolling down a slope. The mass of the cylinder is 122 g and its radius is 5.0 cm. It starts at the top, 23 cm above the bottom, with zero velocity. Use energy considerations.
(a) What is the moment of inertia of the cylinder about its center?
(b) What is the angular speed of the cylinder at the bottom of the slope?
(c) What is the linear speed of the cylinder at the bottom?

P7.5. A wheel of fortune begins spinning with an angular acceleration of 3.1 rad/s^2. What will be its final angular speed after it has rotated 1 revolution?

P7.6. A wheel of diameter 13 cm is spun with a force of 8.2 N on its rim. Its moment of inertia 0.42 kgm^2.
(a) Find its angular speed after 12 s.
(b) Find the angle through which it has traveled in 12 s.

P7.7. A bicycle is racing on a circular track of diameter 66 m. One lap takes the cyclist 45 s. Find the cyclist's average angular and linear speeds.

P7.8. The earth has a radius of 6.38×10^6 m.
(a) What is the linear velocity of a person at rest at the equator?
(b) At what latitude (angle with the equator) would the person's speed be 1/3 that on the equator?

P7.9. The moment of inertia about an axis that is located at a distance ℓ from its center of mass is $I_{cm} + m\ell^2$, where m is the mass of the object and I_{cm} is its moment of inertia about its center of mass. Find the moment of inertia of a solid wheel about an axis through its rim.

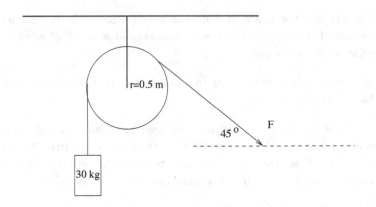

Fig. 7.12   Box of mass 30 kg being raised.

P7.10. A pulley of radius 0.50 m is being used to raise a 30.0 kg mass, as shown in Fig. 7.12. The force is applied at an angle of 45° with respect to the horizontal. What torque about the center of the pulley is required?

P7.11. A meter stick of negligible mass is balanced on a fulcrum located at the 30 cm mark. If a mass of 1 kg is suspended at the 10 cm mark, what mass must be used at the 80 cm mark to balance the stick?

P7.12. (Challenge problem) A uniform plank 3.0 m long is leaning against a smooth wall. The plank has a mass of 15 kg and makes an angle of 60° relative to the ground. (Hint: You need to analyze the torques about the point of contact with the ground).
(a) Find the push of the wall on the ladder.
(b) What is the normal force of the ground on the ladder?
(c) What must be the static coefficient of friction so that the ladder does not slip?

P7.13. (Challenge Problem) When you are in a car that loops the loop on a vertical circular track of 9.1 m radius, and don't want to lose contact at the top of the loop, what is the minimum speed the car can have at the bottom of the loop? (Use energy consideration).

P7.14. A uniform plank of mass 15 kg and length of 5.0 m is placed on a dock with 1.5 m protruding over the edge. How far can a child of 23 kg move beyond the edge of the dock without falling in?

P7.15. Calculate the speed of a solid rolling cylinder (mass m and radius R) at the bottom of an incline of length L and height h if it started at rest at the top of the incline. Use energy principles to solve.

P7.16. A 3.1×10^3 kg space vehicle is sent aloft to get close-up pictures of the moon. The radius of its path around the moon is 3.1×10^6 m. The mass of the moon is 7.38×10^{22} kg. Neglect the effect of the earth. It is desired to move the spaceship closer to the moon.
(a) Should the satellite speed increase or decrease? Why?
(b) What force is needed to get the satellite to orbit the moon at a radius of 2.8×10^6 m from 3.1×10^6 m?
(c) What will be the speed of the satellite at the new radius?

P7.17. From Eq. (7.20), find the torque $\Delta L/\Delta t$ if both I and ω vary.

P7.18. A penny is at the rim of a 12 inch diameter record which is turning at $\omega = 45$ rpm.
(a) What is the penny's tangential speed (in m/s)?
(b) What is its tangential acceleration (in m/s^2)?
(c) What is its centripetal acceleration (in m/s^2)?

P7.19. Calculate the centripetal acceleration of a person on earth (a) at the Equator (b) at latitude 45 degrees.

P7.20. A biker is traveling at a speed of 25 mph.
(a) Calculate the rotation rate of the bike's 0.66 m diameter wheels.
(b) What is the centripetal acceleration of a point on a wheel's rim?
(c) What is the instantaneous velocity of a point on the top of a wheel?
(d) What is the instantaneous velocity of a point on the bottom of a wheel?

P7.21. Estimate your centripetal acceleration as your bus, car or bike turns a corner.

Answers to odd-numbered questions

Q1. Yes, if two equal and opposite forces act on the body at different places; yes.

Q3. No. The moment of inertia of the larger wheel increases as the radius R^2.

Q5. It will slow down.

Q7. At the two poles.

Q9. It will read less.

Q11. The speed will increase.

Q13. The solid cylinder because it has a smaller moment of inertia.

Answers to odd-numbered problems

P1. $f = 1/T = 1/0.35 = 2.86$ Hz. $\omega = 2\pi f = 18/\text{sec}$.

P3. Time for return $= 4 \times 10^4$ m$/3 \times 10^8$ m/s $= 1.33 \times 10^{-4}$ s; $4° = 1/90$ revolution; $\omega = 0.011$ rev$/(1.33 \times 10^{-4}$ s$) = 84$ revs/s.

P5. $\theta = \frac{1}{2}\alpha t^2$; $t^2 = 4\pi/3.1$ rad/s$^2 = 4.05$; $t = 2.0$, and $\omega = \alpha t = 3 \times 2.0 = 6.0$ rad/s.

P7. $\omega = 2\pi/45$ s $= 0.14$ rad/s; $vwr = 33$ m $\times 0.14 = 4.6$ ms.

P9. $I = (3/2)mr^2$.

P11. 50 cm $\times W = 1$ kg $\times 20$ cm; $W = 2/5 = 0.4$ kg.

P13. At the top of the loop, the centripetal force is supplied by gravity, $mv^2/R = mg$. The kinetic + potential energies are $2mgR + \frac{1}{2}mv^2 = 2.5mgR$. At the bottom, there is only kinetic energy $= \frac{1}{2}mv'^2$; thus $v'^2 = 5gR$, and $v' = \sqrt{5gR}$.

P15. $K = \frac{1}{2}mv^2 + \frac{1}{2}I\omega^2 = mgh$; $I = \frac{1}{2}mr^2$; $(3/4)v^2 = gh$; $v = \sqrt{\frac{4}{3}gh}$.

P17. $\tau = \langle I\rangle \Delta\omega/\Delta t + \langle\omega\rangle\Delta I/\Delta t$.

P19. $R = 6380$ km $= 6.38 \times 10^6$ m; $\omega = 2\pi/8.64 \times 10^4 = 7.27 \times 10^{-5}$ (sec)$^{-1}$.

(a) At the Equator, $a_c = R\omega^2 = 3.37 \times 10^{-2}$ m/s^2.

(b) At latitude 45 degrees, the radius of rotation (the distance from the rotation axis) is $R_{\text{Equator}} \times \cos(45°) = 4.51 \times 10^6$ m; $a_c = 2,38 \times 10^{-2}$ m/s^2.

P21. Assume a mass of 70 kg, a car speed of 20 mph $= 8.94$ m/s and a radius $R = 2$ m. Then $a_c = mv^2/R = 70$ kg $\times (8.94$ m/s$)^2/2$ m $= 2.8 \times 10^3$ m/s^2.

Chapter 8

Solar System and Gravitation

8.1 Earth-centered and Sun-centered Theories

Ancient civilizations valued the study of astronomy because it was important to their welfare. The cycle of the seasons for planting and gathering, the periods of the Moon and tides, and the return of longer days after the winter solstice, were significant events.

To anyone watching the skies, it seems that the Sun, the Moon and the stars circulate around us; that Earth is the center, that the universe is *geocentric*. Yet a belief that the sun is the center ... *heliocentrism* ... is in Sanskrit writing of 9th's century BC India. Yajnavalkya wrote that the Sun is "the center of the spheres". He recognized that the Sun is much larger than Earth, and he invented a method for measuring the relative distances of the Moon and the Sun from Earth. The first Greek heliocentrist on record was Aristarchus, in the 3rd century BC, and he too estimated the distances to the Moon and the Sun, apparently by the same technique as Yajnavalkya's (see Fig. 1.2).

Heliocentrism and the implication that the celestial objects were composed of physical substances were ridiculed by Plato and Aristotle. Plato claimed that the perpetual motion and their "obvious" circulation about the earth proved these objects to be divine. His proof of the immortality of the soul begins, "Every soul is immortal, for whatever is in perpetual motion is immortal." He warned that truth comes more from the "eye of the soul than the eye of the body". Truth was to be found through *teleology*: the principle that everything that happens is done for a definite objective, and the entire cosmos has been planned and set in its proper order by a superior force or principle: *Mind*. After Plato, Aristotle described the stars as celestial fires being carried around by the *aether*. Aether is composed of a perfect substance, *quintessence*, the fifth essence after Earth, Water, Air and Fire. Aristotle criticized Aristarchus and other observers of nature who failed to argue from the teleological principle. Aristotle's teaching was accepted for sixty generations, and it was enshrined into the teachings of the world's three great religions. Aristotle's teaching was held in such reverence that it clamped a cold hand of dogma on the study of the physical universe, until the beginnings of the Renaissance in the 15th century.

Yet geocentrists had a problem: not all stars are carried around together; observers knew from the earliest times that certain stars wandered in irregular motions through the field of the fixed stars. These *planets* sometimes moved faster, and sometimes in retrograde motion, opposite to the advance of their neighboring stars. Eudoxus in the 5th century BC invented a solution to *save the phenomenon* of heavenly motion as perfectly and steadily circular. He proposed that each planet is carried by a system of nested spheres, that revolve at different rates, with axes tilted in different directions, and all of them so coordinated that the composition of their motions carries each planet in its complicated path. Eudoxus required 26 spheres to "explain" Mercury's motion. His model dominated the theory of the cosmos until it was found to be inadequate, due to the improving accuracy of observation. It was then supplanted by a theory of *epicycles*,

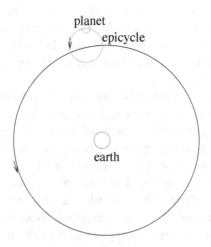

Fig. 8.1   Ptolemy's epicycles-circles rolling along circular hidden tracks.

which reached its final formulation in the work of Claudius Ptolemy of Alexandria, in the second century AD. In this model, the planet is carried around in a circular orbit whose center is carried around on another circular orbit, and so on, until there were enough epicycles with different diameters and speeds that could be adjusted to fit the observed motion. See Fig. 8.1.[a]

8.2   Copernicus and Heliocentrism

The theory of epicycles began to be discarded when it was discovered that the motion of the inner planets Mercury and Venus could be most simply described as a *single epicycle centered on the Sun*. And then, in a complete break with geocentrism, Nicolas Copernicus, a Polish astronomer, in *de Revolutionibus Coelestibus, published 1543*, proposed *heliocentric* (sun-centered) orbits for *ALL* of the planets, including Earth. He wrote,

Finally we shall place the Sun himself at the center of the Universe. All this is suggested by the systematic procession of events and the harmony of the whole Universe, if only we face the facts, as they say, "with both eyes open".[b]

---

[a]The Search for Exoplanets. A great discovery in astronomy at the beginning of the third millennium is that there are planets around other stars — lots of planets around lots of stars. There are three ways that planets may be detected around distant stars. (All require favorable orientation.)
(1) The star moves back and forth along the line of sight as the planet orbits, and the star's periodic motion causes a periodic Doppler shift of the star's light.
(2) The passage of the planet in front of the star causes a temporary dimming of the star's light.
(3) *Microlensing*; a planet's gravitation can focus the light from a distant star, causing a brief brightening. But is there life? That is a much more difficult question to answer. Life as we know it won't be on HD 189733b in the constellation Vulpecula 63 light years distant from us. A "hot-Jupiter" planet close to a star a bit smaller than our sun, HD 189733b is too hot and too massive for life. Nevertheless, the detection of methane and water in its atmosphere, reported in an issue of *Nature* in 2008, is an important step in the search. Not that the presence of these gases tells us anything about life, rather it demonstrates the sensitivity of the technique. As the planet passes in front of the star the absorption spectrum (physics which is treated in a later chapter) of the light grazing the rim of the planet gives us its atmosphere's composition. The measurement stretched the capability of Hubble's near infrared spectrometer to the limit. The search should be much easier with the more powerful James Webb Space Telescope, scheduled for launch in 2013.
[b]Long years before, Aristarchus had proposed that the Earth moved about the Sun, but the Aristotelian view became the accepted theory. By the time that Copernicus entered the scene, Aristarchus' pioneering idea had been forgotten.

Copernicus' model gave a simple explanation for a planets retrograde motion; when the planet's orbit carried it between Earth and the stars, because Earth in its orbit had a greater angular velocity. Copernicus was supremely confident of his analysis:

If there should chance to be any mathematicians who, ignorant in mathematics yet pretending to skill in that science, should dare, upon the authority of some passage of Scripture wrested to their purpose, to condemn and censure my hypothesis, I value them not, and scorn their inconsiderate judgment.

Galileo, who in his early days was a geocentrist, came to believe in Copernicus' theory when he turned his new invention, a telescope, on the sky. He could resolve the bright planet Jupiter as a disk, and he saw that it had attendant points of light that circulated around it as the planet traveled in its complicated path. They were Jupiter's moons! Galileo took Jupiter and its satellites as models of the Sun and its planets; they were proof to him that the Earth could be ... indeed, actually is ... a satellite of the Sun. But aside from his greatness as a scientist, Galileo was a businessman and a self publicist. (He ran a boarding house, and he sold telescopes that he had made). He used his discovery of Jupiter's moons to gain favor and support from the powerful Medici family, by naming the moons, "The Medician Stars".

But heliocentrism could be dangerous, for it contradicted Church dogma. Giordano Bruno had been a Copernican zealot, who asserted that it could be a centerpiece for an all embracing philosophy. He was judged guilty of heresy by the Inquisition in 1600, and burned at the stake. Galileo was more diplomatic, but in 1633 he too came afoul of the Inquisition. He was judged guilty of heresy, placed under house arrest and forbidden to publish. (Although he surreptitiously arranged for the publication of "Two New Sciences" in Holland). Galileo's works were proscribed by the Church, and remained banned until the middle of the 19th century. The Vatican officially rehabilitated Galileo in 1992.

A play *Life of Galileo* by Bertholt Brecht represents Galileo's trial as the conflict between dogmatism and science.

8.3   Brahe and Kepler

The next major player in the study of planets was Tycho Brahe (1473–1543), a Danish nobleman. He was lecturing on astronomy at the University of Copenhagen in the late 16th century, when he received a grant of funds from King Fredericke II to build an observatory. He had it built it on an island near Copenhagen and then began a lifetime of observations of planets' positions. Using instruments of his own design, he made measurements much more precise than any that had been done before; some to an angular precision of 0.01°, contrasted to the 0.3° of earlier measurements. His work became known; he received a letter from Johann Kepler, asking if he could come and inspect his data, so as to "perfect his theory". Kepler had some reputation at that time, owing to his book on the planets. His work was sheer mysticism and numerology. Questions such as, "Why should there be just six planets? Was there any significance to a certain succession of numbers that approximated their distances?" He tried relating them to nested geometrical figures, inscribing and circumscribing triangles, squares, hexagons and other figures. Euclid had shown that there were only five regular solids (whose faces were all absolutely alike); Kepler decided that they corresponded to the gaps between the planets. He went on to the planets' motions, with comparable numerology. He tried to imagine some propelling force emanating from the sun. When he received Kepler's letter, Brahe welcomed him, and he offered Kepler the post of scientific assistant. And that is when Kepler began working in earnest, on Brahe's extensive data.[c]

Sir Oliver Lodge, a historian of science, wrote, "It is difficult to imagine a stronger contrast between two men ... one (Brahe), rich, noble, vigorous, passionate, strong in mechanical ingenuity and experimental skill, but not above the average in theoretical and mathematical power. The other (Kepler), poor, sickly, devoid of experimental gifts, and unfitted by nature for accurate observation, but strong almost beyond competition in speculative subtlety and innate mathematical perception." We can add one more distinguishing feature of Brahe: after the end of his nose had been sliced off in a duel, he wore a silver nose.

Working with Brahe's precise measurements, Kepler derived three *Laws* of planetary motion. In 1609 he published the first two, and in 1619 the third. From the time he had begun to puzzle over the planetary orbits until he published the third law, it was twenty two years.

---

[c]Here we must note the irony, that although Brahe's measurements were the data on which Kepler based his theory of planetary orbits about the Sun, Brahe himself was a geocentrist. He rejected motion of the Earth as contradicting the Bible and Aristotelian physics.

Johannes Kepler

Tycho Brahe

A 1610 portrait of Johannes Kepler
unknown artist

Fig. 8.2   Pictures of Brahe and Kepler, courtesy of Wikipedia.

Kepler's Laws

1. *The orbits of the planets are ellipses,*[d] *with the Sun at one focus of the ellipse.* (See Fig. 8.3).
2. *The areas swept out in unit time by the radius vectors of the planets are a constant of the motion.*
3. *The cubes of the planets' average distances from the Sun are proportional to the squares of their orbital periods.*

Such mathematical precision raised questions about the fundamental mechanism. The first question was: *What force could hold the planets in their orbits about the Sun?* For they did have to be held, otherwise they

---

[d]An ellipse is an oval, like a squashed circle. But it is a more precise figure. You can draw an ellipse by placing two thumbtacks on a board, and encircling them with a loop of string that is longer than twice their spacing. Then run a pencil around the loop, pressing the pencil against the string as you go around. The pencil will trace out an ellipse. The thumbtacks' positions are the foci of the ellipse.

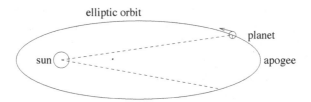

Fig. 8.3   Kepler's elliptic planetary orbit with sun as one focus.

would fly off in straight trajectories, just as a stone, whirled about your head at the end of a string, goes flying out when the string breaks. Could the planets be held in their orbits by gravity?

8.4   Newton and Universal Gravitation

Here is an excerpt from notes that Newton wrote when he was seventy-three: *In that same year (1666) I began to think of gravity extending to the orb of the moon, and having found out how to estimate the force with which a globe revolving within a sphere presses the surface of the sphere, from Kepler's rule of the periodical times of the planets being in a sesquialterate proportion[e] of their distances from the centers of their orbs. I deduced that the forces which keep the planets in their orbs must be reciprocally as the squares of their distances from the centers about which they revolve; and thereby compared the force requisite to keep the moon in her orb with the force of gravity at the surface of the earth, and found them answer pretty nearly. All this was in the plague years of 1665 and 1666, for in those days I was in the prime of my age for invention, and minded mathematics and philosophy more than at any time since.*

In fact, Newton was not alone in conjecturing about gravitational force. His contemporary Robert Hooke proposed several ideas about motion and gravitation independently. He proposed that the force varies as the inverse square of the distance, even earlier than Newton, which enraged Newton. When Hooke read an early version of the *Principia* he complained that his several contributions were slighted. Then Newton, in fury, went systematically through the manuscript, removing every mention of Hooke's name and work.

---

[e]Sesquialterate means one and a half times, or that the squares of the periods vary as the cubes of the orbital radii.

The next and more intricate question was *Could an inverse square law explain elliptical orbits?* Newton's answer, after a long and hard calculation, was *Yes!* The mathematics (calculus) is beyond our level, but it has become a staple of university physics courses.

The theory of gravitation has a much greater scope — it is *UNIVERSAL: Every bit of matter attracts every other bit.* The force depends on their masses as well as their separation. This is the "Law of Universal Gravitation",

$$F = \frac{Gm_1m_2}{r^2} \tag{8.1}$$

Look at its simplicity: Each mass is there, m_1 and m_2, multiplying each other. Therefore you and the Earth attract each other, and *you exert as much force on the Earth as it does on you.*

Illustrative Question: By how much does the force change for an object if its distance from the earth's center is doubled?

Answer: Since the force is inversely proportional to the square of the separation, the force becomes one quarter of its original value.

Illustrative Problem: Determine g from Eq. (8.1) in terms of the radius and mass of the earth.

Answer: $F = mg = mM_EG/r_E^2$. Therefore $g = M_EG/r_E^2$. Check this out!

The distance r between you and the Earth presents a problem. Newton guessed that the attraction is as if the Earth's mass were concentrated at its center, so for an object on the surface of the Earth the proper value for r is the Earth's radius. He later proved the guess right (assuming the Earth's mass to be spherically symmetric); it needed his new mathematics, the calculus.

The factor $G = 6.67 \times 10^{-11}(\text{Nm}^2)/\text{kg}^2$ is the *universal Gravitational Constant*. It was measured first by Henry Cavendish in a laboratory experiment. It has since been determined with much more sensitive techniques. An ongoing study at the University of Washington is measuring G with much greater precision, yet, and also examining the exactness of the inverse square law.

Kepler's 3rd law is readily proved from Eq. (8.1), in the simplifying case of circular orbits. Balancing the centrifugal force against gravitational attraction, the condition for a planet's orbital frequency ω, mass m and distance R from the Sun (of mass M) is

$$m\omega^2 R = GmM_s/R^2$$

Combining factors of R and writing $\omega = 2\pi/T$, we get

$$R^3/T^2 = GM_s/4\pi^2 . \tag{8.2}$$

The case of elliptical orbits takes fancier mathematics. Gleick's biography tells that Newton undertook the proof after being questioned by the astronomer Edmund Halley; "What sort of orbits would be consistent with an inverse square law?" "Ellipses", said Newton, claiming that he had calculated that result years before. But he could not immediately produce the proof. He then began a deeper study into the implications of the inverse square law, one of the most intense periods of his life.

Another application of angular momentum conservation is in Kepler's laws of planetary motion. Although there is a force acting on each planet, its moment arm is away from the sun to the planet, so that the $\vec{r} \times \vec{F} = 0$, since $\sin\theta = 0$, and there is no resulting torque. The force is called a *central force* and the angular momentum of a planet is constant. On the scale of planetary motion, the planet and sun can be considered to be "point particles". The moment of inertia of the planet is thus simply mR^2, where m is its mass and R is its distance from the sun. As this distance decreases, so does its moment of inertia. Thus ω has to increase to compensate and keep the angular momentum constant. The area of any small segment of motion is proportional to R^2 so that equal areas are traversed in equal times; this follows from $I\omega \propto R^2\omega = $ constant. This is one of Kepler's three laws. You can do a related experiment at home. If you twirl a ball at the end of a string in a circle, then, as you shorten the string you will find that the ball goes faster.

8.5   Tides and Phases of the Moon

The Earth's tides are due to the Earth–Moon gravitational attraction. Figure 8.4 illustrates that the force is stronger on the side of the Earth facing the Moon, because the distance is less, so the oceans' waters tend to mound up on that side. As the Earth rotates, the mound continues to face the

Fig. 8.4   Earth's tides.

Moon, so at fixed positions on Earth the tide rises and falls once each day. A second, weaker tide occurs midway between these surges. It is due to the fact that the Earth and the Moon form a gravitational couple, rotating about their center of mass (see Fig. 8.4). The parts of the Earth that lie furthest from the Moon experience the greatest inertial reaction, i.e. the greatest centrifugal force, hence the waters on that side mound up. You may think of it as the Moon pulling the Earth away from the water. Thus there is a second tide each day, about 12 hours apart from the first one, one on the side facing the Moon, and one on the opposite side. The Sun's gravity adds to the effect, when Sun and Moon are lined up on the same side of the Earth; then there is an especially high "high tide". Some locations on Earth experience weaker or stronger tides than others, due to local peculiarities of the ocean bottom and the shoreline. In some parts of the South Pacific, the tidal height is only a few centimeters, while at the Bay of Fundy in the State of Maine, a resonance effect produces tidal heights greater than 16 meters (!!).

The plasticity (squishiness) of the Earth allows tides in its solid parts as well as the oceans. The energy due to the tidal action causes internal heating which, together with the radioactivity of some elements, raises the temperature of the core above the melting point of the rock. These are the sources of heat for the lava plumes that sometimes come to the surface in volcanoes.[f]

What causes the *phases* of the Moon that we experience each month? The Moon reflects sunlight and this is how we see it. We can understand the phases of the Moon by considering this feature. The Moon is full when

---

[f]You can demonstrate the warming effect of 'internal friction' by kneading bakery dough or modeling clay.

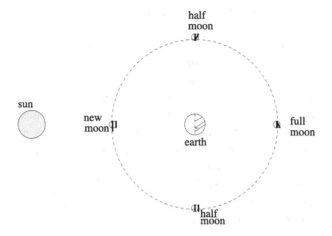

Fig. 8.5 Phases of the moon — the dark lines and cross-hatching indicate shadow.

it is on the opposite of the Earth from the sun. The Earth rotates about its own axis on a daily basis, and the Moon rotates about the Earth in 27.3 days. The new Moon occurs when the Moon and Sun are on the same side of the Earth. We see a half Moon half-way between these two extremes and this is close to the time when we can see the Moon and the Sun at the same time. A crescent Moon occurs when we are beyond or before a half-Moon. See Fig. 8.5.

A *lunar eclipse* occurs when the earth casts its shadow on the moon and a *solar eclipse* happens when the reverse occurs: the moon casts its shadow on the earth.

8.6 Satellites

Many artificial satellites (and much satellite debris) are circulating about the Earth. They serve a variety of missions, including communications, Earth observations, astronomical research, and the Global Positioning System, GPS. The value of satellites for astronomical observations is due to their altitude, far above clouds and atmospheric haze. Experiments in Space Lab are examining the behavior of plants, people and physical phenomena at "micro-g" conditions.

8.7   Gravitational Potential Energy

If you raise an object a distance from h_1 to h_2, you increase its gravitational potential energy Δ(P.E.) $= mg(h_2 - h_1)$. This assumes that the weight of the object is constant, independent of height, which requires that h_1 and h_2 be much smaller than R_E, the Earth's radius. But if the distance is comparable to or larger than R_E, the increase is less, because the gravitational force decreases with distance. The *gravitational force varies as $1/R^2$, but the gravitational potential energy varies as $1/R$.* (A proof needs calculus, but we state it without proof here.) It is convenient to set its value to zero at infinite distance; then it becomes progressively more negative, i.e. "deeper", as R decreases, since the potential energy decreases as we get closer and closer to the Earth.

$$(P.E.)_R = \frac{-Gm_1m_2}{R}\,. \qquad (8.3)$$

Now, a few applications of Eq. (8.3). Meteorites fall to earth from remote regions of the solar system. As they fall, their gravitational potential energy is gradually converted into kinetic energy. Assume the meteorite's mass is m and the Earth's mass is M; if it falls from very far (essentially, infinity) to a distance R, the change in gravitational potential energy is

$$\Delta(P.E.) = -GMm/R\,. \qquad (8.4)$$

The lost P.E. is converted into kinetic energy.

Illustrative Example: How fast is a meteorite traveling when it hits the Earth?

Answer: Let's assume that it was at rest when it was infinitely far away. Since the meteorite had zero velocity at an infinite distance, its kinetic energy and potential energy were both zero, so its total energy is zero. As it approaches Earth it loses potential energy; at a distance R its potential energy is: $(P.E.) = -GMm/R$. Since total energy is conserved, *total energy* $= 0 = mv^2/2 - GMm/R$. Then at $R = R_E$, the speed when it hits the Earth is $v = \sqrt{2GM/R_E}$. Note that we're neglecting the effect of air resistance. Actually, the frictional heating is so great that small meteorites burn up and never reach the ground.

8.8 Dark Energy

The Universe is expanding with an ever increasing rate, due to something the astronomers call *dark energy* that is pushing it apart faster than gravity can be pulling it closer. Until about five billion years ago the expansion was slowing due to gravity between the atoms and the *dark matter* that make up about one quarter of the Universe. But as the expansion continued the influence of dark energy, which is thought to make up the rest of the universe, began to surpass that of gravity, and the rate of expansion speeded up. No one claims to understand dark energy.

8.9 Antigravity

There's no such thing, but people have long dreamed of inventing a machine, a method, or a magic spell to cancel gravity. However, it lives in science fiction and the comics. Superman turns it on and off at will; if you look carefully at scenes where he's flying with someone under his arm, it seems that he's also able to switch gravity off on that person too.

In "Cyrano de Bergerac", a play by Edmond Rostand, the hero Cyrano proposes several techniques for countering gravity. One: Stand on an iron plate, and throw a magnet overhead; then, before it falls back, throw another one up; etc, etc. Two: Recognizing that the Sun draws vapors up from water, put some water into a jar, seal it closed, and grasp it tightly while standing in bright sunlight.

We can suggest a method for anyone with a pet cat. You know that when a cat falls it always lands on its feet. And a slice of buttered bread always falls with the buttered side down. Now, just tape a buttered slice of bread, buttered side up, on your cat's back, and then throw the cat up in the air.

But there are a few real situations where the local gravitational force is zero.

1. Between two gravitating bodies, where their attractions to a third mass are equal and opposite. See Problem 5.
2. In an orbiting satellite, such as Spacelab.
3. During a "suborbital flight" in a plane that makes a looping, centripetal trajectory. A couple of commercial organizations are now planning to sell space for passengers on such flights. Virgin Atlantic has booked more than 200 tickets, at $200,000 each, for 5 minute excursions. A few years

ago a research plane made a series of such flights to study the effects of zero gravity on people. It was called "The Vomit Comet".

Table of Astronomical Masses

Earth	5.98×10^{24} kg
Sun	1.99×10^{30} kg
Moon	7.35×10^{22} kg

Another useful table could be planets' periods and distances. You can get them on the Web.

Questions

Q8.1. Astronauts in a satellite experience a "micro gravity" environment. What effects might prevent it from being exactly "zero-g"?

Q8.2. There is an ongoing search for extrasolar planets, i.e. planets belonging to stars other than our sun. How can the existence of a planet be detected?

Q8.3. Why is it advantageous to launch a rocket from a low latitude location (near the Equator), rather than a location closer to the poles? And which way would you launch it, N, S, W, E?

Q8.4. What is the torque exerted on the Earth (or other planet) as it rotates about the sun? Give a reason.

Q8.5. Where does a planet move fastest as it orbits the sun? Where does it move slowest? Explain.

Q8.6. When during the day or night would you expect to see the new moon rise and set? Explain.

Q8.7. Why is there a high tide when the Moon is on the opposite side of the Earth from the ocean? The pull of the Moon on the water is the weakest here.

Q8.8. Do Kepler's three laws apply to satellites orbiting the earth? Explain.

Problems

P8.1. Consider the range of a projectile shot horizontally from a raised platform. The curve of the Earth increases it, so that it is greater than the range on a flat plain. Calculate the minimum velocity for the range to equal the Earth's circumference. Assume that the projectile can skim the surface of a smooth perfectly spherical Earth with its mean radius, and no atmosphere.

P8.2. A geostationary satellite will orbit at the same rate as the Earth's rotation, so that it "hovers" above one location on the Equator.
(a) Calculate the altitude (height above the Earth's surface; $R_E = 6.38 \times 10^6$ m) of a geostationary satellite, assuming "standard" g, i.e. the same value as at the Earth's surface. (This is, of course, unrealistic.)
(b) From your answer in (a), calculate how much the value of g would actually differ from the standard value.

P8.3. The "Vomit Comet" airplane is used for studies of zero-g effects. It executes "looping" flights, so that at the top of the circle its downward acceleration is equal to g. Calculate the algebraic relationship between the radius of the circle and the plane's speed that combine to produce zero-g at the top of the loop.

P8.4. Pluto is so far from the Sun, 6.0×10^9 km, that its proper designation as a planet has been revoked in 2006. Calculate its orbital period, in "Earth years". Suggestion: Try working the problem algebraically as a ratio, T(Pluto)/T(Earth).

P8.5. Calculate the position (distance from the center of the Earth) where the Earth's and Moon's attractions on a mass m cancel. The data you need for the calculation are: Earth's mass 5.98×10^{24} kg, Moon's mass 7.35×10^{22} kg, Earth–Moon distance (between centers) 3.84×10^8 m.

P8.6. Challenge. This problem will illustrate the "retrograde motion" of some planets. Draw a diagram of the Earth's orbit and the orbit of Mars. From Kepler's Laws, calculate the angular velocity of a planet as a function of its distance from the Sun. Show how the greater *angular* velocity of the Earth makes it appear that Mars can sometimes seem to be moving backwards.

P8.7. How hard must you throw a stone to escape completely from the grip of the Earth's gravitational pull (assuming no air resistance)?

P8.8. Compare the pull of the Sun and Moon on the Earth. The Sun's mass is 1.99×10^{30} kg and Moon's mass is 7.35×10^{22} kg. Their average distance from the Earth are 1.50×10^{11} m and 3.84×10^8 m, respectively.
(a) Find the average forces and their ratio.
(b) What is the impact of the Moon on the Earth's orbit about the Sun?
(c) What is the effect of the sun on the moon's path if you use the Earth-Sun distance as the Moon-Sun distance?

P8.9. Mary has a mass of 55.0 kg on the surface of the Earth. What would her mass and weight be if she doubled her distance from the center of the Earth?

P.8.10. The period of the Moon about the Earth is 27.3 days.
(a) What is the fraction of the Earth's period about the Sun for one period of the Moon's motion about the Earth?
(b) After one period of the Moon, will it be in the same position relative to the Earth? If not, why not?
(c) Show that it takes an extra 2 days (29.3 days) for the Moon to appear full again.

Answers to odd-numbered questions

Q1. The satellite might be decelerating, due to frictional drag in the very thin gas of the upper stratosphere. And there is a very slight gravitational attraction of the spaceship itself.

Q3. Launching toward the East would give the satellite a boost from the Earth's rotational velocity. The boost would be greatest from an equatorial site.

Q5. The Earth (or planet) moves fastest when it is closest to the Sun and slowest when it is furthest from the Sun. The force is a maximum when the planet is closest to the Sun.

Q7. There is a high tide because the Moon pulls the Earth away from the water and this pull is a maximum here.

Answers to odd-numbered problems

P1. The centripetal acceleration would have to be equal to g, the acceleration of gravity. Thus, $v^2/R = g$; with $R = 6.38 \times 10^6$ m, $v = 7.91$ km/s. Note that, since the projectile would encircle the Earth, it could continue for any number of orbits; thus, its range would be infinite.

P3. Algebraically, $v_{\min} = \sqrt{(Rg)}$.

P5. Let the total Earth–Moon distance be L_o, the distance between the mass m and the Earth be L_E, and the distance between the mass m and the Moon be L_M. Then, since gravitational attraction varies as m/d^2, the condition for equilibrium is $M_E/M_M = (L_E/L_M)^2$. Substituting $L_M = L_o - L_E$, after algebra we get $L_E = L_o/[M_E \pm \sqrt{M_E M_M}]/(M_E - M_M) \approx 0.890 L_o = 3.4 \times 10^8$ m. Note that only the "–" sign of the *pm* makes sense.

P7. The stone's initial kinetic energy would have to be greater than the increase of potential energy in traveling from the Earth's surface to an infinite distance away. Thus, $mv_o^2/2 = GM_E m/R$, where $R = 6.38 \times 10^6$ m is the Earth's radius. Since the stone's mass cancels from both sides of the equation, $v_o = \sqrt{2GM_E/R} = 1.12 \times 10^4$ m/s.

P9. Mary's mass would remain 55 kg. The gravitational force is 1/4 that on the Earth. Her weight would thus be 135 N.

Chapter 9

Oscillations and Waves

9.1 Oscillatory Motion — Elastic Spring

Oscillatory motion is repetitive motion. An example is the simple pendulum discussed in Chapter 5. Such motion can be described mathematically by a repetitive function, e.g., sine or cosine.

An example of oscillatory motion is that of an elastic spring. We can suspend a mass from the spring and stretch it to begin oscillatory motion, but it may be easier to think of a spring on a frictionless table, so that gravity is not involved. When we stretch the spring (or compress it) we are adding *elastic potential energy* to it. However, we have an example here of a non-constant force. When a spring is stretched, there is a force of restitution; that is the spring wants to get back to its equilibrium position. The more you stretch the spring, the bigger this restitutional force. The force satisfies *Hooke's law* if the stretch is small,

$$F = -kx \,, \tag{9.1}$$

where x is measured from the equilibrium position and k is a measure of the spring's stiffness; it is called the *spring constant*. The "−" sign indicates that the force is opposite to the displacement. To stretch the spring, we have to apply a force opposite to the restitutional force. The work required to stretch the spring a distance x, if we do not accelerate it, is the average applied force, $kx/2$, times the distance, x, the spring is pulled. Thus, we find

$$\Delta W = \frac{1}{2}kx^2 = \Delta U \,, \tag{9.2}$$

where ΔU is the change of potential energy. The work done goes into the *elastic potential energy* of the spring. The spring can now do work and

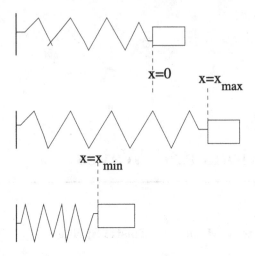

Fig. 9.1   Spring stretch and compression.

if the spring is stretched and let go, it will gain kinetic energy as it loses potential energy, but the total mechanical energy is constant if we neglect friction. The spring will attain its maximum speed and kinetic energy when it crosses the equilibrium position, but the motion will keep going and the spring will be compressed by the same amount it was stretched. Because mechanical energy is conserved

$$K_{\max} = \frac{1}{2}mv_{\max}^2 = U_{\max} = \frac{1}{2}kx_{\max}^2 \,, \qquad (9.3)$$

where m is the mass attached to the spring. The maximum K occurs when $x = 0$, where we can choose to set $U = 0$, whereas the maximum U occurs when $x = x_{\max}$.

The total mechanical energy

$$E = K + U = constant \qquad (9.4)$$

The motion is oscillatory or "periodic" with the equation

$$x = x_{\max} \cos(\omega t) \,, \qquad (9.5)$$

if $x = x_{\max}$ at $t = 0$, and $x_{\max}$ is the amplitude of the oscillatory motion. It is a cosine wave, because at $t = 0$, the spring has been stretched and $x = x_{\max}$. Note that ωt is an angle; it is called the *phase* of the oscillatory motion. If the time taken for a complete oscillation is T (called the *period*), then $\omega T = 2\pi$ radians, or $\omega = 2\pi/T$.

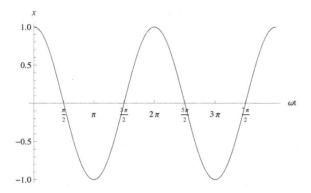

Fig. 9.2   Plot of $x = x_{\max} \cos(\omega t)$, with $x_{\max}=1$.

But what determines ω or T? Newton's second law tells us that $F = ma$; thus we need to solve the equation of motion for the case of a time-varying force. We have not yet learned how to do this, but the solution is Eq. (9.5), with

$$\omega = \sqrt{\frac{k}{m}}, \tag{9.6}$$

$$f = \frac{\omega}{2\pi} = \frac{1}{2\pi}\sqrt{\frac{k}{m}}, \tag{9.7}$$

$$T = \frac{1}{f} = \frac{2\pi}{\omega} = 2\pi\sqrt{\frac{m}{k}}. \tag{9.8}$$

Frequencies, f, can be given as cycles/s, expressed as Hertz or Hz. The motion is called *simple harmonic*. The *frequency* of oscillation is $f = 1/T = \omega/(2\pi)$. Mechanical energy goes from only potential energy to a mix of kinetic and potential energies to purely kinetic energy and so forth.

It is worthwhile to note some of the features of Eq. (9.5) or of a cosine curve. The slope (speed) is a maximum when the cosine goes upward through zero and is zero at the maximum and minimum of the curve. The slope is always positive when the *cosine* increases and always negative when it decreases; see Fig. 9.2.

Illustrative Problem: A spring stretches 6.0 cm when a mass of 0.6 kg is suspended from it.
(a) Find the spring constant.

(b) If the mass is set into oscillations, determine the period and frequency of oscillation. (Forget gravity).

Solution: (a) $F = mg - kx = 0$ at equilibrium. Thus $k = \frac{9.8 \times 0.6 \text{ N}}{0.06 \text{ m}} = 98$ N/m.

(b) $T = 2\pi\sqrt{(m/k)} = 2\pi\sqrt{0.6kg/98 \text{ N/m}} = 2\pi \times 0.0775 = 0.49$ s.

9.2   Harmonic Motion of a Pendulum

Galileo was fascinated by the swinging motion of a hanging lantern; it motivated him to begin the study of the oscillations of a gravitational pendulum. We have seen the pendulum in Chapter'5. The motion of the bob describes an arc of a circle (see Fig. 9.3). If we lift the mass and let it go, then

$$\theta = \theta_{\max} \cos(\omega t) , \tag{9.9}$$

where $\theta_{\max}$ is the maximum angle and we assume the pendulum starts its motion there from rest. The motion is oscillatory and there is a continual exchange of potential and kinetic energies.

What determines the constant ω? Angles are measured in radians (see Chapter 7) and ωt is an angle. We can write Eq. (9.8) as

$$\theta = \theta_{\max} \cos \frac{2\pi t}{T} , \tag{9.10}$$

since $\omega = 2\pi/T$. If the maximum angle is small then we have approximately $\sin\theta \approx \theta$ and the net restoring force (towards the equilibrium position at the bottom of the swing) is

$$F = -mg\cos(90° - \theta) = -mg\sin\theta \approx -mg\theta . \tag{9.11}$$

This is not a constant force; it is reminiscent of the force on a mass tied to a spring. Here the displacement $x = L\sin\theta \approx L\theta$ (see Fig. 9.3), so that the restoring force can be written as $F = -\frac{mg}{L}x$; thus the spring constant k is replaced by mg/L. For the spring we saw that $\omega = \sqrt{\frac{k}{m}}$ and this becomes

$$\omega = \sqrt{\frac{mg}{mL}} = \sqrt{\frac{g}{L}} , \tag{9.12}$$

so that

$$f = \frac{\omega}{2\pi} = \frac{1}{2\pi}\sqrt{\frac{g}{L}} , \tag{9.13}$$

$$T = \frac{2\pi}{\omega} = \frac{1}{f} = 2\pi\sqrt{\frac{L}{g}} . \tag{9.14}$$

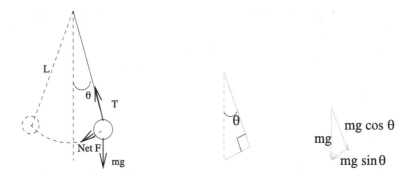

Fig. 9.3   Motion and forces on the bob of a pendulum.

From the spring and the pendulum, we see that the requirements for oscillatory motion is a restoring force proportional to the displacement and low damping. We have assumed no friction so far, but it is ever present. If the rate of loss is small, the frequency is not greatly affected, but the amplitude decreases *exponentially*, so that Eq. (9.8) is modified to

$$\theta = \theta_{\max} e^{-\kappa t} \cos(\omega t) , \qquad (9.15)$$

where κ is the *damping constant*. If κ is large, oscillatory motion may not occur.

Illustrative Problem: A pendulum 1 m long is pulled through a small angle of 0.1 rad and set into oscillation. The mass of the bob is 100 g. Find (a) The tension in the suspending string, and (b) the oscillating frequency ω. Is the tension constant for larger angular displacements?

Solution: (a) The forces must balance perpendicular to the motion of the bob, i.e. along the string. Thus $T = mg \cos \theta \approx 0.1 \text{ kg} \times 9.8 \text{ m/s}^2 \times 1 \approx 1 \text{ N}$. (b) The circular frequency $\omega = \sqrt{\frac{g}{L}} = \sqrt{\frac{9.8 \text{ m/s}^2}{1 \text{ m}}} \approx 3.1 \text{ rad/s}$. The tension is not constant for larger angles.

A child on a playground swing is applying a principle of physics to increase the arc of her motion. With no one to push her, she begins to rock back and forth, and gradually excites the swing into larger and larger amplitudes. She senses when to rock, to *resonate* with the natural frequency of the swing.

Less happily, you are carrying a bowl of soup from the kitchen to the dinner table, and the motion of your body resonates with the sloshing of the liquid. Making matters worse, when you attempt to stop the sloshing, your timing is off, and ... !!! ("Don't blame me; it was a resonance!")

9.3 Waves

We are all familiar with water waves in the ocean or on a lake. A large wave can knock you over, which suggests that waves carry both energy and momentum. In order to have a mechanical wave, the requirements are similar to those for oscillations, but an extended, deformable medium is also necessary.

There are many different kinds of waves. The simplest kind of wave is a regular, repetitive one, called a *periodic wave*. Sound can be an example of a repetitive wave and so is light from the sun. We shall study these two in more detail. A pulse sent down a string, by means of moving the string perpendicular to the string itself, is also a wave, but not a repetitive one. If you tie a taut string to a post and move the other end up and down, you again can get a wave. If the up and down motion is repetitive, you get a periodic wave.

Waves have some distinguishing properties from objects or particles. One of them is that two objects cannot be in the same place at the same time, but two waves can do so. This gives rise to *interference* phenomena, which we shall study.

In waves, a disturbance, such as a compression of the medium, may move in the direction of propagation of the wave, but the disturbance does not travel along with the wave. It may move a small distance in that direction. The disturbance may also move at right angles to the direction of wave motion as for the pulse on a string. Waves can be differentiated by the direction of the disturbance. *Longitudinal* waves are those for which the disturbance is parallel or anti-parallel to the direction of propagation of the wave. An example is the compression and rarefaction of air for a sound wave. Telephone speakers or receivers have a thin diaphragm and the pressure changes moves the diaphragm back and forth to produce the sound. By contrast, the wave on a string is a *transverse* wave because the disturbance is perpendicular to the direction of motion of the wave. There are also waves which are combinations of these motions.

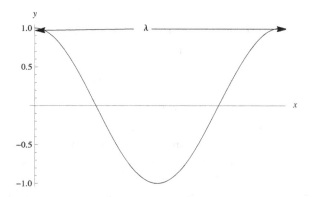

Fig. 9.4   Space dependence of a wave, $y = \cos x$, from $x = 0$ to $x = \lambda$.

In a periodic wave, the time between repetitions is called the *period*, just as for oscillatory motion. The number of repetitions per unit time is called the *frequency*, so that

$$f = 1/T \,. \tag{9.16}$$

Unlike an oscillation on a spring, waves tend to travel and propagate in a certain direction. We can look at the wave as it passes by us as a function of time, as in Fig. 9.2 and can write the time dependence as in Eq. (9.5). If the wave is transverse and propagates in the x-direction

$$y = A \cos 2\pi \frac{t}{T} \,, \tag{9.17}$$

where A is the amplitude of the wave. Alternatively, we can take a snapshot of a wave at a fixed time, and describe what we see as (see Fig. 9.4)

$$y = A \cos 2\pi \frac{x}{\lambda} \,, \tag{9.18}$$

where, again, we have assumed that the wave travels in the x-direction, but that the disturbance is perpendicular to the direction of propagation, along y. See Fig. 9.4. We have also assumed that at $x = t = 0$, $y = A$. This is what gives rise to a cosine wave.

If we combine Eqs. (9.16) and (9.17), we obtain

$$y = A \cos \left[2\pi \left(\frac{x}{\lambda} - \frac{t}{T} \right) \right] \,. \tag{9.19}$$

The quantity inside the square bracket is called the *phase* of the wave. The equation gives us the velocity of the wave in the x-direction by seeking how x and t change for a constant phase

$$v = x/t = \lambda/T = f\lambda. \tag{9.20}$$

The velocity of a wave is related to its frequency and wavelength, $v = f\lambda$. Two waves are said to be *in phase* if they reach their maxima and minima at the same time. (This assumes that their wavelengths are the same). They are said to be *in phase opposition* or to interfere destructively if one reaches its maximum when the other one reaches its minimum. They are out of phase, but not in phase opposition, for any other situation.

What is the physics that determines the velocity of a wave? It depends on the properties of the medium that it travels through. For a wave on a rope, it also depends on the *tension, $\mathcal{T}$*, of the rope, which provides the accelerating force on the neighboring pieces of the rope. When a piece of rope is moved up and down, this piece is connected to its neighbor, which also feels the pull. We assume the rope to be uniform and very long; thus it is the mass per unit length, $\mu = m/L$, which determines the velocity

$$v = \sqrt{\mathcal{T}/\mu}, \tag{9.21}$$

where $\mathcal{T}$ is the tension in the rope. The larger μ is, the larger the inertia. For a sound wave in a gas (e.g. air), the speed is determined by the masses of the molecules, as well as the temperature and pressure of the gas. In a liquid or solid, the molecules are closer together, so that you would expect a faster sound velocity, which is the case. In air, at normal pressure and temperature, the speed of sound is approximately 340 m/s, whereas in iron it is 1500 m/s and in water it is 1480 m/s. In a gas of hydrogen, which is about 7 times lighter than air at the same pressure and temperature, we expect the speed to be larger than in air, and it is, 1300 m/s.

The range of frequencies that are audible to the human ear is approximately 5 Hz to 15 kHz. Elephants hear lower frequencies and dogs can hear higher ones. Bats use ultrasonic sound (frequencies higher than those heard by individuals) to navigate, and locate obstacles, food, and prey. They use the reflected sound, or *echo*. *Sonar* works on the same principle. A series of regularly spaced sound pulses and their reflection is used as a depth gauge and to search for underwater objects.

We have not mentioned light or radio waves; their speed of propagation is $c = 3 \times 10^8$ m/s in vacuum and (approximately) in air. Light waves do *not* need a medium for propagation. (It is the electric and magnetic fields that do the oscillating).

9.4 Standing Waves

When a wave on a rope hits the retaining nail or other retainer at the end, what happens? The wave pulls on the nail. By Newton's 3rd law there is an equal and opposite force that the nail exerts on the rope. This causes a reflection of the wave. If the transverse wave was increasing when it reached the end, it pulls the nail up and the nail then pushes down on the rope. The reflected wave is thus of opposite phase (180° or *out of phase*) to the incident one (see Fig. 9.4). At any one place on the rope there are now two waves, one moving to the right, say the incident wave, and one moving to the left, the reflected wave. Both waves have the same frequency and wavelength. These waves *interfere*.

If the two waves of the same amplitude reach their maxima at the same time and in the same place, then we have *constructive interference*, and the amplitude of the resulting wave doubles. If the two waves arrive at a certain position out of phase, one being a maximum, when the other one is a minimum, then we have *destructive interference*, the resulting wave vanishes at that location, and the amplitude is 0. Since for an incident and reflected wave, the propagation is in opposite directions, there will be places with constructive and places with destructive interference. We get what is called a *standing wave* with *nodes* (minima) and *antinodes* (maxima) if there is a *resonance*.

As an example, consider a wave on a rope of length L, attached to a wall at one end and nailed down at the other end. If the middle of this rope is plucked in a repetitive manner, we get an oscillatory wave. If the reflected wave reaches a maximum at the hand (at $L/2$) just as the latter passes through its maximum, then the two waves are in phase at the hand, and the amplitude increases. An apt analogy is pushing a child on a swing, when the push comes at regular intervals just as the swing reaches its maximum. The time for one cycle is T, and the time for the wave to return to the far end of the rope is $2L/v$. If $T = 1/f = 2L/v$ the reflected and original waves reinforce. The two waves are said to be *in resonance*. The above condition is $f = v/2L$ and $\lambda = v/f = 2L$. The frequency and wavelength for this condition are called the *natural* or *fundamental* frequency and wavelength. We can get a different frequency if the string is plucked

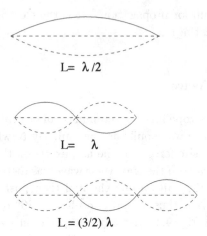

$$L = \lambda/2$$

$$L = \lambda$$

$$L = (3/2)\,\lambda$$

Fig. 9.5   Some examples of standing waves on a rope tied down at both ends.

at $\frac{1}{4}L$ and $\frac{3}{4}L$ at regular intervals, as in Fig. 9.5(b). That is, we have a resonance whenever

$$f = n\frac{v}{2L}, \tag{9.22}$$

$$\lambda = 2L/n, \tag{9.23}$$

where n is an integer. The longest wavelength or smallest frequency is the natural one. The waves are called *standing waves* because the maxima (*antinodes*) and zeros (*nodes*) do not propagate. The lowest frequency is called the *fundamental mode*, and the higher ones are called *overtones*. The overtones are called *harmonics* if they are multiples of the fundamental frequency. Some examples are shown in Fig. 9.5. Stringed musical instruments, e.g., violins, guitars, etc. use such standing waves.

When a string is plucked at an off-center position, it will begin to oscillate at the fundamental and several harmonic frequencies. The higher frequencies tend to die away more rapidly than the fundamental, so that sound evolves from a complex note with *overtones* to a deeper, simpler one. Stringed instruments use this bit of physics. The bow is coated with a sticky wax, *rosin*. When the bow is drawn over the instrument, there is a strong viscous frictional drag. If the bow is pulled faster than the wax can flow, the force increases until the wax bond suddenly yields, and the force is reduced. But the bow is still moving, the wax again seizes the instrument's strings, and the process repeats. This stick-slip motion excites the

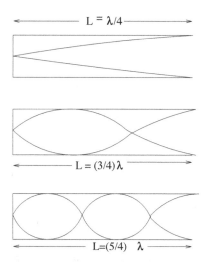

Fig. 9.6 The first three harmonics for an organ pipe open at one end.

musical string into its resonant tone and several overtones. If the overtones and harmonics are in the ratio of integers, the sound is pleasant. Stick-slip motion is also what causes chalk to squeak on a blackboard and a wine glass to ring resonantly when you rub a moist finger around the rim.

If the rope is tied down at one end and "waved" at the other end, the natural wavelength is $4L$, or only $1/4$ of the wave fits on the rope. Smaller wavelengths that resonate are $L = 3\lambda/4$, and quite generally

$$(2n + 1)\lambda/4 = L. \tag{9.24}$$

Examples are shown in Fig. 9.6. An organ pipe uses this principle if it is open at one end and closed at the other one. The open end is where sound comes out. The other end is where sound is reflected. If this end is closed, it is like the end of a string or rope with a nail. The amplitude of the wave needs to be zero there (node). On the other hand, for an open pipe, you want the amplitude of the wave to be a maximum there to match the openness. The fundamental mode has a wavelength of $\lambda = \frac{1}{4}L$, where L is the length of the organ pipe. For an organ pipe open at both ends, the longest, or fundamental, wavelength is $\lambda = 2L$. See Fig. 9.7.

Sound waves have many uses. Ultrasound waves are used in medicine. You send sound into the body and wait for the reflection or echo. An example is an echo cardiogram, which reflects from the heart of a patient. You can observe changes in densities, malignancies, and more benign growths,

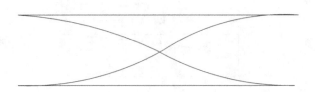

Fig. 9.7   The fundamental mode for an organ pipe open at both ends.

such as babies. The higher the frequency of the instrument, the shorter the wavelength and the finer the details you can observe. These instruments are a useful diagnostic tool.

Illustrative Problem: A musical reed instrument is an oboe. Assume that it is open at one end. Take the length of the oboe to be 1.0 m. Determine the two lowest resonant frequencies if the speed of sound is taken as 330 m/s.

Answer: The longest wavelength in resonance is $4L = 4.0$ m. This corresponds to a frequency of $v/\lambda = 330$ m/s/4 m = 83 Hz. The next resonant wavelength is $4/3L$, versus the longest one of $4L$, so that the frequency is v/λ is three times as high as the fundamental = 250 Hz.

9.5   The Doppler Effect for Sound — Echoes

When a train is approaching someone at an intersection, its whistle gets shriller. Its sound waves are at a higher frequency than that for the train standing still. This effect of motion on the perceived frequency is called the *Doppler effect*. To understand it, consider the case of an approaching source of sound. As the source approaches, the wave crests get closer, so that the wavelength is reduced by $v_s T$, where v_s is the speed of the source. Thus the perceived wavelength is

$$\lambda' = \lambda - v_s T \tag{9.25}$$

and the perceived frequency is

$$f' = \frac{v}{\lambda'} = \frac{v}{\lambda - v_s T} = \frac{v}{\lambda(1 - v_s/v)} = \frac{f}{1 - v_s/v}, \tag{9.26}$$

where v is the sound velocity. When the train recedes, the opposite happens — the apparent frequency is lower,

$$f' = \frac{f}{1 + v_s/v} \,. \tag{9.27}$$

When the observer moves, a similar effect occurs, but the reason is somewhat different. As you move towards a source at speed v_o, you encounter more crests per unit time than if you were standing still. The extra number you encounter in a time t is $v_o t/\lambda$. The wavelength does not change. For one period, this number is $v_o T/\lambda = v_o/v$. Thus, the frequency f' that you hear is

$$f' = f(1 + v_o/v) \,. \tag{9.28}$$

You may have noticed the Doppler shift when a fire engine or a police car passes, with its sirens blaring. The Doppler effect is used by the police to detect speeders and by tennis pros to measure the speed of a serve. The Doppler effect is also responsible for the *red shift* of distant galaxies.

When a sound wave encounters a smooth wall, one with irregularities small compared to the wavelength, it is reflected like light from a mirror. The reflected sound is sometimes called an *echo*. Bats use these echoes to locate food or obstacles. The size, shape, and composition of an auditorium determines its echoing qualities or reverberations. A hard surface yields echoes whereas a soft and porous one tends to absorb the sound.

9.6 Beats

In the section on standing waves, we discussed the interference of two waves in space. Waves can also interfere in time. Consider sound being emitted by two tuning forks placed near each other, with slightly different frequencies or periods. Since the sound waves emitted by the forks travel at the same velocity, they will exhibit slightly different wavelengths. The two waves interfere and will produce a somewhat complex wave pattern, as illustrated in Fig. 9.8. The superposition of the two waves results in a pattern that has, in addition to the two waves, an increase and decrease in sound amplitude (and therefore intensity) at frequencies that correspond to the difference and sum of the two primary frequencies. It is the difference in frequencies that is easier to detect by the human ear, especially if the individual frequencies are quite high. As an example of beats, if the tuning forks are at 10 and 11 Hz, the lower *beat* frequency will be at 1 Hz, as illustrated in Fig. 9.8.

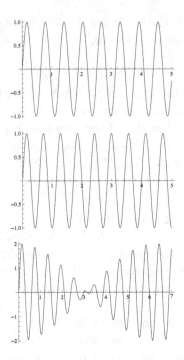

Fig. 9.8 Superposition of two waves differing by 10% in wavelengths. (a) and (b) are two pure sine waves, and (c) the resulting wave when they are superposed.

Questions

Q9.1. Give some examples of elastic potential energy other than those given in the text.

Q9.2. Will any restoring force give rise to simple harmonic motion? Explain.

Q9.3. In the pendulum, when is K a maximum? When is the potential energy a maximum? Is mechanical energy conserved if we neglect friction? Give reasons for your answers.

Q9.4. List all the forms of mechanical energy involved in a bungee jump. Explain.

Q9.5. What is the wavelength of a sound wave of 300 Hz in air? Take the speed of sound in air as 340 m/s.

Q9.6. In an oscillating spring, at what positions in the motion is the speed a maximum and where is it zero? Repeat for the acceleration.

Q9.7. Is the speed or frequency that a rope is waved up and down related to the wave speed? If so, how and if not, why not?

Q9.8. If you use a tuning fork with a known frequency, how can you use an organ pipe and standing waves to measure the speed of sound in air? Explain.

Q9.9. Can you produce a standing wave on a rope fixed at both ends? If so, how? If not, why not?

Q9.10. Why does the velocity of sound in air depend on the temperature?

Q9.11. Can sound travel in vacuum? Explain.

Q9.12. A spring is suspended from a hook and has a mass of 100 g attached to it. What is the effect of gravity on its equilibrium position and on its oscillatory motion? Explain.

Q9.13. While singing in the stall shower, you notice that a particular note echoes more strongly than others. Why?

Q9.14. Will a pendulum clock run faster or slower in hot weather? Why?

Problems

P9.1. When the spring is at $1/2$ of its maximum stretch, is the potential energy $1/2$ of its maximum value? Find K and U at $x = \frac{1}{2}x_{\max}$ in terms of k and $x_{\max}$.

P9.2. Relate the wavelength of a standing wave on a rope tied down at one end to those of the incident and reflected ones.

P9.3. For the cases of open and closed organ pipes determine the next two lowest frequencies, after the fundamental, in terms of the pipe length.

P9.4. A string instrument has a wire of length $L = 68$ cm between two fixed points. The wire has a mass/length $= 0.0016$ kg/m and is under a tension of 140.0 N.
(a) What is the velocity of a wave on this string?

(b) What is the wavelength of the fundamental note? What is its frequency?
(c) Draw this standing wave and that of the next higher harmonic. What is its frequency?
(d) Determine the number of nodes and antinodes for the 3rd harmonic.

P9.5. An organ pipe of length 1.4 m is closed at one end. What are the fundamental wavelength and frequency?

P9.6. (a) A strung bow, when pulled back, acts like a harmonic oscillator. If a certain bow has a spring constant of 472 N/m and is pulled back for a distance of 0.53 m, what is its elastic potential energy?

(b) An arrow of mass = 0.025 kg is added to the bow, when pulled back horizontally as in part (a). When let go, what is the horizontal velocity of the arrow?

P9.7. A pendulum has a length of 0.84 m. If pulled back through a small angle and let go, how long will it take to reach its equilibrium position?

P9.8. A spring is suspended from the ceiling and a mass of 520 g is added. When the mass is added to the spring, the equilibrium position drops by 12 cm.
(a) Determine the spring constant.
(b) How will gravity affect the oscillation of the mass and spring?
(c) If the spring is pulled a further 8.0 cm, what is the elastic potential energy due to the pull? What is the total potential energy relative to the displaced equilibrium position?
(d) If the mass is let go, determine its frequency of oscillation.
(e) Determine the kinetic energy of the mass as it passes the displaced equilibrium position.

P9.9. There is a rule of thumb that states that you can tell the distance of a thunderstorm in miles by taking the time delay in seconds of the sound of thunder to reach you (relative to the reception of the lightning) and dividing it by 5.
(a) Explain how and why this works.
(b) If the time delay is 8 s, determine the distance in km? (5 miles $\approx$ 8 km).

P9.10. (more difficult) A policeman in a squad car suddenly comes behind you as you are traveling 40.0 mph in a 30.0 mph zone. He turns on his siren as he travels the same speed as you.

(a) Will you hear the same frequency as that emitted by the police siren? Explain.

(b) If at a certain time, the frequency emitted by the moving police car is 400.0 Hz, what frequency will you hear if you have stopped, but the policeman is still traveling at 30.0 mph?

(c) What frequency would you hear before you slowed down and the policeman is traveling at 30.0 mph?

P9.11. A bat in a large cave flies towards a small opening at a speed of 22 m/s. The bat receives the echo from the surroundings of the opening. If the emitted sound is 2000 Hz, determine the frequency that the bat will hear. (The bat keeps moving.)

P9.12. An Austrian alpinist finds herself between two cliffs. When she yodels, she hears 3 echoes. The first one arrives at a time t_1; the second one arrives at $t_2 = t_1 + 1.8s$; the 3rd arrives at time $t_2 + 1.2s$. Neglecting echoes from the ground, she can determine her location relative to the two cliffs and the separation of the cliffs. Do this calculation for her.

P9.13. A bicyclist bikes toward a jazz festival and hears the sound at a frequency that is 1% higher than the festival patrons. What is the bicyclist's speed?

P9.14. (a) A pendulum of length 1.3 m has a bob of 0.52 kg. Determine the period and frequency of the oscillation.

(b) What length should you choose for the pendulum to have a period of 1 s, so that it can function as a clock?

(c) What is the tension in the suspension of the bob?

P9.15. A piano is tuned. The tension for middle C is 956 N in a wire that is 0.690 m long and is tied down at both ends. Find the linear density (mass/length) of the wire if the period of middle C on this wire is 3.59 ms. Assume that middle C is the natural frequency.

Answers to odd-numbered questions

Q1. Vibrating membrane (drum), bow and arrow, rubber band, etc.

Q3. K is a maximum at the center and P.E. is a maximum at the maximum angle. Mechanical energy is conserved if friction can be neglected.

Q5. $\lambda = v/f = \frac{340 \ m/s}{300/s} = 1.13$ m.

Q7. No. The rate of waving is related to the frequency.

Q9. Pluck the center at appropriate intervals.

Q11. No. It needs a medium to compress and decompress.

Q13. This note resonates inside the confined chamber of the shower walls.

Answers to odd-numbered problems

P1. No. $\frac{1}{2}kx^2$ is only 1/4 of the maximum when $x = 1/2$ of its maximum value and thus the kinetic energy, $K = \frac{3}{4}kx^2_{\max}$.

P3. For an open organ pipe, the wavelengths are found from $L = \lambda/2$, λ, $3\lambda/2$, so that the next to lowest frequencies are v/L, and $3v/(2L)$; for a closed pipe, the wavelengths are found from $L = \lambda/4$, $L = 3\lambda/4$, $L = 5\lambda/4$. Thus the two next higher frequencies are $3v/(4L)$ and $5v/(4L)$.

P5. $\lambda = 4L$ and $f = v/(4L) = 340$ m/s$/4 \times 1.4$ m $= 61$ Hz.

P7. $t = T/4 = \frac{\pi}{2}\sqrt{\frac{L}{g}} = \frac{\pi}{2}\sqrt{0.84/9.8} = 0.46$ s.

P9. (a) Because the time delay depends linearly on the distance.
(b) 8 s/5 miles = (8/5 miles)(8/5 km/mile) = 64/25 = 2.6 km.

P11. The signal going out from the bat has the source going toward the receiver (cave wall); thus $f'/f = \frac{1}{1-v_s/v}$, where v_s is the speed of the source (bat). The echo comes from a stationary source and propagates towards a moving receiver (bat); thus $f'' = f'(1 + v_o/v)$. Thus the frequency heard by the bat is $f''/f = \frac{1}{1-v_{\text{bat}}/v}(1 + v_{\text{bat}}/v) = (1 + 22$ m/s$/340$ m/s$)/(1 - 22/340) = 1.14$; thus $f'' = 2280$ Hz.

P13. $f' = 1.01f$; $f'/f = 1 + v_o/v$; $f'/f - 1 = 0.01 = v_o/v$, or $v_o = 0.01v = 3.4$ m/s.

P15. $v = \sqrt{T/\mu}$; $f = v/\lambda$, and the period $T' = \lambda/v = \lambda/\sqrt{T/\mu}$. For the fundamental $\lambda = 2L$. Thus, $\mu = T'^2T/(4L^2) = 6.52 \times 10^{-3}$ kg/m.

Chapter 10

Thermal Physics

Fire is one of the ancient Greeks' "elements". They valued it for its transforming qualities; winning metals from their ores and later annealing them, cooking food, heating houses, and lighting dark places. But it had its dangers; small cooking fires sometimes became uncontrolled conflagrations that could consume whole cities. "Oh fire", said Pliny, "thou measureless and implacable portion of Nature, shall we rightly call thee destroyer or creator?"

10.1 Heat is a Form of Energy

The fire's essence that gives it these powers is *heat*. But what is that? Until the late 18th century heat was believed to be a material substance, the *caloric fluid*. It flowed into bodies when they were heated and flowed out when they cooled. Friction caused bodies to get warmer because the rubbing surfaces were pressed together, which squeezed out the caloric fluid. The theory seemed reasonable until the experiments of one of the most colorful figures in the history of science.

Benjamin Thompson was born in a farming community in Massachusetts in 1753. He became, in various periods of his life, country gentleman, major of the New Hampshire militia, royalist spy, diplomat for the American colonies, knight in England, a colonel in the British army, and founder of the Royal Institution of Great Britain. He wrote to a friend in 1785, "*I can say with truth that I hardly know what there is for me to wish for. Rank, Titles, Decorations, literary distinctions, and with some small degree of military fame I have acquired, the road is open to me for the rest.*" There was indeed one more. Working as an undercover agent for

the British government, he offered his services as aide-de-camp to the ruler
of Bavaria, which was one of the states of the Holy Roman Empire. On
May 9, 1792, a proclamation was read:

*We, Carl Theodor ... pleased with the excellent personal merits of Sir
Benjamin Thompson ... promote him to the rank and dignity of the Imperial
Counts of the Holy Roman Empire, with the title of Count of Rumford.* And
through it all, Thompson maintained a keen interest in science. He invented
a thermometer, and with it investigated heat conduction and convection in
liquids. His most important work was described in a paper published in
1798, a study of the heat generated by the friction of boring machines
in the manufacture of cannons. He proved that the heat produced was
inexhaustible, and directly proportional to the depth of the bore. Then
he compared the thermal properties of the chips (their heat capacity) with
the metal of the cannon, and found them to be identical — there was no
evidence that any substance, such as a caloric fluid, had been squeezed out
of them. He concluded that:

> Anything which any insulated body ... can continue to furnish without
> limitation, cannot possible be a material substance, and it appears to me
> to be extremely difficult, if not quite impossible, to form any distinct idea
> of anything, capable of being excited and communicated in the manner
> the Heat was excited and communicated in these experiments, except it be
> MOTION.

By that, he later explained, he meant the heat absorbed by a body
excites internal motion of the material. It was a notion that was remarkable
for its time; it was confirmed by experiments more than two centuries later.

Thompson's proof that heat is a form of energy was quantified by the
Englishman James Prescott Joule, who was also a brewer. He experimented
with various types of mechanical energy that generated heat. The culmina-
tion of his years of work was reported in a detailed paper *On the Mechanical
Equivalent of Heat*, published in 1850. He showed that mechanical energy
is convertible to heat. The apparatus was simple in principle. The mechanical energy is produced by the falling of weights; which rotate a paddle that
stirs a quantity of water in an insulated container, and the internal friction
(the viscosity) of the water causes a temperature rise. By these means the
mechanical work is directly compared to the unit of heat, the *calorie* (defined below, a few paragraphs later). The importance of his study made
his name immortal: in the metric system the unit of work is the *Joule*.

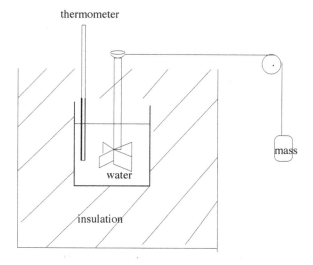

Fig. 10.1   Joule's proof that heat can be obtained from mechanical energy.

10.2   The First Law of Thermodynamics

Thermodynamics is the study of heat. Thompson's research is summarized by the *The First Law of Thermodynamics*, which states simply that *energy is conserved with heat taken into account*. Explicitly, the first law distinguishes three terms: ΔQ, the heat *absorbed* by a substance, ΔW, the work done *by* the substance, and ΔU, the change in its internal energy,

$$\Delta U = \Delta Q - \Delta W. \tag{10.1}$$

Note the signs: ΔQ is positive if the body gains heat, and ΔW is positive when work is done *by* the system. The internal energy of a body is the sum of the kinetic and potential energies of the individual atoms that comprise the system. In some ionizing gases, ΔU also includes chemical and electrical energies. At this point it is not necessary to specify atomic details — the total internal energy ΔU will suffice.

In Fig. 10.2 we sketch an example of the breaking of the first law. In the figure, water is made to come out from the bottom of a tank and is brought back up to the top through a pipe. At the top, it is used to turn a fan or a turbine. A farmer tried to patent the device. Even neglecting friction, it could not work because there is no energy left when the water gets to the top.

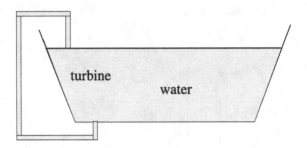

Fig. 10.2   Example of the breaking of the first law of thermodynamics.

10.3   Temperature

We must distinguish between temperature and heat.

Heat is a form of energy that "flows" toward lower temperature. It can be converted from mechanical, chemical or other forms of energy, but its distinctive character is its flow from hotter to colder regions.

Temperature describes a state of the body; a measure of "hotness". In a body at thermal equilibrium, every part of the body, no matter how small, has the same temperature. However, if a body is heated locally, temperature is no longer uniform. When the body is removed from the source of heat, the heat flows toward the regions of lower temperature, until the temperature becomes uniform throughout, and equilibrium is reestablished.

Temperature can be measured by a variety of *thermometers*, devices with properties that vary with temperature in a regular way. A familiar type consists of mercury or alcohol in a sealed glass tube. Since the liquid expands with temperature more than the glass, the height of the liquid column rises when the temperature increases. The thermometer is *calibrated* by measurements at standard temperatures, such as the freezing and boiling points of water at atmospheric pressure. Another common type is a *bimetallic strip*; two dissimilar metals joined together along their length. The metals have different expansion coefficients, so when the strip is warmed one side expands more than the other, causing the strip to bend. Bimetallic strips are used in dial thermometers, and thermostats that control room temperature.

Temperature is specified on a temperature scale. Two scales in common use are the Celsius (t_C) and Fahrenheit (t_F) scales. For the two scales, the freezing point of water is $0°C = 32°F$ and the boiling point of water is

$100°C = 212°F$. Temperature readings can be converted from one scale to the other by the formula

$$t_F = (9/5)t_C + 32 \,. \tag{10.2}$$

If you take a gas in a fixed container (constant volume), and lower its temperature (on the Celsius scale), its pressure (force/area — see Chapter 11) drops. A plot of pressure times volume versus temperature for a perfect gas (see Chapter 11) is shown in Fig. 10.3. A remarkable feature emerges. When the line through the experimental points is extended, it reaches zero at $PV = 0$ and $T = -273.2°C$. What does this mean? It is the lowest temperature possible. This temperature is called 0, on the *Kelvin, or Absolute Temperature scale*; it is written $T = 0$ K. The absolute temperature is written with no superscripts, and in place of the degree sign °, just the letter K. The Kelvin scale is simply related to the Celsius scale by:

$$T = t_c + 273.2 \,. \tag{10.3}$$

No one has ever reached 0 K; the lowest temperature that has been achieved is $\sim 10^{-7}$ K. Classically, at 0 K nothing moves — e.g. electrons, atoms, etc. Note that the temperature interval, one degree, is the same on the Celsius and Kelvin scales. The thermodynamic basis for the absolute temperature scale is based on a proof devised by a young French engineer in the 19th century, Sadi Carnot.

A great engineer was Sadi Carnot;
Proved Absolute Temp. begins at Zero.
But fame did escape him; did History fail,
For Kelvin's the name that we give to the scale.

What is the highest temperature possible? In principle there is no limit. The sun's temperature is of the order of 10^7 K and the temperature at the beginning of the universe was of the order of 10^{29} K.

10.4   Heat Capacity

One of the primary thermal properties of a substance is its *heat capacity*. The specific heat capacity (usually shortened to *specific heat*) of a substance is the amount of heat required to raise the temperature of one unit of mass one degree. In common use, specific heat is measured in calories per gram per Celsius degree,

$$c_H = \Delta Q/(m\Delta t) \quad \text{cal/g}°\text{C} \,. \tag{10.4}$$

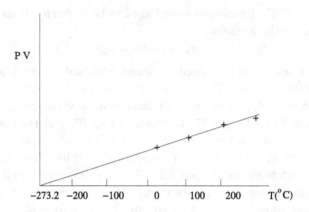

Fig. 10.3   Plot of pressure times volume versus temperature (°C) for a perfect gas.

Which brings us to the definition of the calorie: *A calorie is the quantity of heat that raises the temperature of 1 gram of water from 20°C to 21°C at normal atmospheric pressure.* The mechanical equivalent of heat J was first measured by Joule. The modern value is

$$1 \text{ calorie} = 4.184 \text{ J}. \tag{10.5}$$

By definition, the specific heat of water at 20°C is 1.00 cal/(g°C); that of ice is only 0.49 cal/(gm-°C). The specific heat of steel is 0.11 cal/(gm-°C).

The specific heat of water is quite a bit larger than that of most substances. For example, the specific heats of glass, minerals, and metals at room temperature are, typically, a few tenths that of water. Even ice has a specific heat which is only about half that of water. Water's large specific heat has a major moderating effect on the environment: regions near the ocean coasts and along the borders of large bodies of water tend to have milder climates than inland regions; less hot in summer and less cold in winter.

Illustrative Problem: A bottle of wine of mass = 500 g, is at a temperature of 20°C. The specific heat of the bottle and wine is 0.2 cal/g-°C. It is immersed in a 1 kg bath of water at 5°C. If 300 cal are given up by the bottle to the water,

(a) determine the change in internal energy of the wine bottle in Joules.

(b) Find the temperature change of the water and of the wine.

(c) At what temperature will equilibrium be reached and what is the exchange of heat energy to reach that point?

Answer: (a) ΔU (wine) = 300 cal $\times$ 4.19 J/cal = 1257 J $\approx$ 1260 J.
(b) ΔT(water) = $\Delta U/(mc_H)$ = 300 cal/(10^3 g $\times$ 1 cal/g-°C) = 0.3°C. The temperature change of the wine bottle is 300 cal/(500 g $\times$ 0.2 cal/g°C = 3°C.
(c) Equilibrium is reached when the temperatures are the same for the water and bottle; at that point $\Delta Q = \Delta U = 0$, $0.2 \times 500(T - 20) + 1 \times 1000(T - 5) = 0$; $T = 6.4$°C. $\Delta Q = 0.2$ cal/g-°C $\times$ 500 g $(6.4 - 20)$°C = -1360 cal.

10.4.1 *Food energy*

The energy value of foods is given in Calories, with a capital C, where 1 Calorie = 1 kilocalorie. An example is a small candy bar, which has a listed energy content "200 Calories", i.e. 200 kilocalories. Very little of the food energy we consume is used to perform mechanical work; almost all is for maintaining the organism; breathing, pumping blood, and building and repairing cells. A person's *thermal efficiency*, the fraction of heat energy that can be converted to mechanical work, is well below 50%.

10.5 Phase Changes

Does the addition of heat always cause the temperature of a substance to rise? "No" is the answer. In addition to changing an object's temperature, heat can cause a *change of phase*. The common phases of simple materials are solid, liquid, and vapor. Typically, the low temperature phase is a solid; the addition of a *latent heat of fusion* L_f causes the solid to melt to a liquid; further heat raises the temperature of the liquid. At the boiling point a further addition of a *latent heat of vaporization* L_v turns the liquid into a vapor. The temperature of each phase change is slightly sensitive to pressure; if pressure is held constant (as at standard atmospheric pressure), the phase change takes place at a constant temperature.

Some thermal properties of water:

Specific heat (defines the calorie) $c_H = 1.00$ cal/g°C at 20°C

Latent heat of fusion (ice to water) at 0°C, $\quad L_f = 80$ cal/g

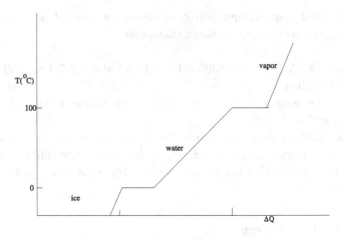

Fig. 10.4   Temperature change of water with the addition of heat.

Latent heat of vaporization (water to vapor) at 100°C, $L_v = 540$ cal/g

10.6   Heat Flow

The second law of thermodynamics (the study of heat) states that heat tends to "flow" from a hot body to a cooler one, i.e., from a body with a higher temperature to one of lower temperature. There are several mechanisms for heat transfer or flow: conduction, convection, and radiation.

Illustrative Example: How much ice would melt if the bottle of wine in part (a) of the last problem were immersed in 1 kg of ice at 0°C rather than water?

Answer: 300 cal would melt 300 cal/80 cal/g = 3.75 g.

10.6.1   *Heat conduction*

Heat conduction occurs through physical contact. Examples are the heat flow from a pot to its handle, or between a car's tires and the road. The flow of heat is proportional to the *thermal conductivity* of the material, the contact area, and the temperature difference; it is inversely proportional to the length of the conducting path. The rate of heat flow in a solid is proportional to the rate of temperature change with distance: $\Delta T/\Delta x$, the *temperature gradient*

$$\Delta Q/\Delta t = k_T A \Delta T/\Delta x. \tag{10.6}$$

Here Δt is a time interval and the temperature difference is ΔT, A is the cross sectional area of the heat path, and k_T is the thermal conductivity.

There is a great variation of k_T among materials. Pure metals, such as copper and silver, have a high thermal (and electrical) conductivity. Electrical insulators, such as glass, and wood generally have poor heat conductivity. For example k_T(copper) ≈ 400 J/(s $-$ m$^\circ$C); for stainless steel it is ≈ 4 and for glass it is ≈ 0.04 in the same units. These values are all approximate and depend on the purity of the material.

Illustrative Example: The bottom of a stainless steel pot has a diameter of 10 cm and a thickness of 0.5 cm. The pot contains 1 kg of ice in equilibrium with 1 kg of water at 0°C. The pot is heated on a hot plate of the same diameter as the pot and is at a temperature of 50°C. How long will it take to melt the ice if the temperature flow is by conduction. Use k_T(copper) $=$ 400 J/s-m$^\circ$C.

Answer: Using the equation of heat transfer by conduction, we find that it is 310 J/s. The heat of fusion of ice is 80 cal/g $= 334$ J/g. Thus, the time to melt all of the ice is $\approx 9.28 \times 10^2$ s ≈ 15 min.

10.6.2 *Heat convection*

Heat conduction in fluids, liquids and gases, is by *convection* ... Due to thermal expansion, heating a fluid makes it less dense, so it tends to rise. As it rises it loses its heat to the surroundings such as the walls of the container. There it cools, contracts, and sinks. Thus the fluid *convects*; flowing up in the center of the container, and down at the walls, if it is heated from below [see Fig. 10.5(a)]. In a large container the convection may break up into a pattern of upstreaming and downstreaming regions, called convection or *Bénard cells* [see Fig. 10.5(b)]. Evidence of convection cells can be seen sometimes in the natural environment, in the regular spacing of cumulus clouds. The clouds are in the portions of air that are rising and cooling. You can see convection cells in your kitchen, in a shallow pan of cooking oil heated gently on the stove.

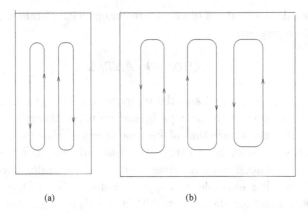

(a) (b)

Fig. 10.5   Heat convection (a) in a small container and (b) in a large chamber.

10.6.3   *Heat radiation*

Radiant heat transfer needs no physical contact; it is the emission and absorption of heat by *electromagnetic waves*. We will study these waves in more detail in Chapter 16. Examples are heat lamps, the Sun, and also the Earth; cosmic background radiation (see Chapter 1) is at 2.725 K. An object absorbs and radiates heat with an efficiency that depends on its color and the character of its surface. The least efficient absorber and radiator is a shiny reflecting surface, like a mirror or a polished pot. On the other hand, a dull black surface absorbs a high fraction of the incident radiation, and reflects little of it. Thus, the most comfortable clothing in the summer are light colored, smooth material. Benjamin Franklin made a simple test: he laid small squares of different cloth, with different colors, side by side on a snow bank. After an interval in the sun, the patches were lying in depressions of varying depths. The depths indicated how much snow had been melted and thus the efficiency of absorption of the sun's rays. The black patch was the deepest.

When an electric heater is turned on, you feel little warming until the heater begins to glow brightly. That is because the rate of heat emission varies with the fourth power of the temperature, as given by the *Stefan–Boltzmann* law.

$$\Delta Q/\Delta t = e\sigma T^4, \tag{10.7}$$

where e is the *emissivity* of the material and σ is a constant.

The *greenhouse effect* has been useful to farmers and horticulturists for centuries; only recently has it come to mean something undesirable. But the physics is the same. In all cases it is a combination of emission, absorption, and reflection of heat waves.

The glass in a farmer's cold frame and that of a botanical garden reflect the radiation from the ground, thereby trapping the internal heat radiation and raising the temperature. In the earth's atmosphere, carbon dioxide (CO_2) and methane (CH_4) have the same effect as the glass by absorbing and reflecting some of the earth's radiation back to the ground. CO_2 is far less effective than methane, but there is a lot more of it. The long term trend to warmer average temperatures correlates with the increase of CO_2 in the atmosphere; the principal sources are the burning of fossil fuels: oil and coal. There is a trend to use renewable energy sources such as wind, solar, and nuclear energy.

Our planet Earth emits black body radiation. Its average temperature is established by an equilibrium between the incoming and outgoing radiation. Not all the energy radiated from the surface escapes; some of it is re-radiated back to the ground. Equilibrium occurs when the rate of emission is equal to the rate of absorption.

10.7 Entropy and Reversibility. The Second Law of Thermodynamics

Entropy is a measure of disorder. Here are some examples.

A deck of cards is perfectly *ordered* when the cards are separated into suits, and numbered in sequence in each suit. When the cards are shuffled, the perfect order is destroyed; the deck's entropy is increased.

A lump of sugar is dropped into a glass of water. As it dissolves and the molecules of sugar disperse into the water, the sugar's entropy increases.

The Second Law of Thermodynamics states: The tendency of all natural processes is to increase the total entropy of the "Universe" (meaning everything that is involved in the process).

When there are several systems interacting, it is possible for one of them to get lower entropy at the expense of the others. For example, here is what happens when you freeze water in a refrigerator. As the water freezes, its entropy is reduced; the atoms in ice are in a regular array, much more orderly than in the liquid. When we put the water in the refrigerator its entropy is lowered, but the refrigerator had to use some electricity to cool it, and the power plant that supplied the electricity had to burn some fuel, and the fuel's entropy was increased. The Second Law says that the net effect was that the total entropy of the Universe was increased.

The Second Law is *statistical*; it pertains to systems with many atoms. In contrast, a very small system can spontaneously order into a state of lower entropy. In the example of the sugar and the glass of water, if there were only a few molecules of sugar and a microscopic drop of water, there is a finite chance that, at some moment, the sugar molecules will all collect into a tiny cluster. For smaller and smaller systems, the fluctuations become more important. Can all molecules in a gas in box move to one side of the box? The answer is yes, but the probability of it happening is vanishingly small unless there are only a few molecules.

For reversible processes there is no change of entropy, but most processes are *irreversible*.

10.8   Heat Engines and Thermodynamic Efficiency

One of the concerns of practical engineers in the 18th century was the efficiency of steam engines, engines that use heat to provide mechanical energy. They wondered, how could one get the most energy out of a fuel? In heating steam to run an engine, would it be more efficient to produce a high temperature under a small boiler, or a lower temperature under a larger boiler? In 1824, Sadi Carnot tackled the question by idealizing the basic principle of a heat engine. He proved that the *ideal thermal efficiency* of an ideal engine (without frictional or heat losses), is directly related to two temperatures: the high temperature T_h of the burning fuel, and the low temperature T_c of the exhaust. (Note that the temperatures are Kelvins.) See Fig. 10.6.

Carnot simplified the description by assuming that the engine operates between a hot reservoir at a temperature T_h and a cold one at temperature T_c. The heat absorbed from the hot reservoir is ΔQ_h and an amount ΔQ_c is rejected at the cold reservoir. The engine produces useful work in the amount ΔW. Conservation of energy gives

$$\Delta W = \Delta Q_h - \Delta Q_c. \qquad (10.8)$$

The thermal efficiency e is related to the ratio of temperatures

$$e = \Delta W/\Delta Q_h = 1 - \Delta Q_c/\Delta Q_h.$$

Carnot engines operating between the same two reservoir temperatures must have the same efficiency, whatever their working substances may be. If this was not so, it would be possible to have a Carnot engine running backwards, as a refrigerator. Due to the difference in efficiency, it is then

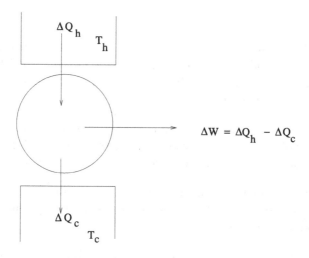

Fig. 10.6   Carnot heat engine.

possible to create more energy than the engine consumed. Thus, it is possible to choose an ideal gas as the working substance, so that

$$\Delta Q_c/\Delta Q_h = T_c/T_h, \tag{10.9}$$

and the efficiency becomes

$$e = 1 - T_c/T_h. \tag{10.10}$$

The equation shows that e is greater for hotter sources and cooler exhausts. An exhaust temperature of 0 K is required for 100% efficiency. For the gasoline engine of a car T_h is the temperature of the ignited fuel in the piston chamber; T_c is the temperature of the exhaust. In practice, a gasoline engine's T_h is limited by the loss of heat through the engine's walls, and T_c is the temperature of outside air. The engine's efficiency is typically not higher than about 0.3. A refrigerator is a heat engine in reverse, as is a heat pump when used to heat a house.

An example of an engine which breaks the 2nd law of thermodynamics is an engine which is promised to be 100% (or close to it) efficient when it operates between normal temperatures such as 300°C and 50°C.

Questions

Q10.1. Heat is an extensive (dependent on mass) quantity, and temperature is an intensive (independent of mass) quantity. Designate which among the following quantities are extensive and which intensive: mass, color, weight, density, volume, hardness.

Q10.2. The specific heat of water is greater than almost every other substance. Explain how this may account for the stability of climate of maritime regions and coastal areas next to large bodies of water.

Q10.3. The volume of a gas increases in a predictable way as the temperature is increased at constant pressure. Could such a system be used as a thermometer? Explain.

Q10.4. You eat a candy bar in preparation for climbing Mt. Rainier. What would your body use the candy for, other than to increase your potential energy?

Q10.5. A lighted match quickly extinguishes during orbits of the Space Lab. Why?

Q10.6. Why does spraying a plant with water protect it from freezing?

Q10.7. Differentiate between "energy conservation" in common use and the conservation of energy in physics? Find at least one way in which you could change your daily activities to improve your "energy conservation".

Q10.8. Why is it advantageous to cool a bottle of wine in a bath of water in equilibrium with ice rather than in ice, alone?

Q10.9. Why do the handlebars of your bike feel colder than the rubber grips in winter?

Q10.10. A double pane window consists of two panes of glass, with a sealed space of dry air between them. The window has a high thermal resistance, which it loses when a leak develops into the space between the panes. Explain.

Q10.11. Can you add heat to an object without raising its temperature? Explain.

Q10.12. Do you need contact with a hotter body in order to raise the temperature of a substance?

Problems

P10.1. Estimate the cost of a 10.0 minute shower from the following: water temperature = 60.0°C, flow = 5.0 liters/minute. The initial temperature of the water entering the water heater was 40.0°C; it was heated electrically at a cost of 5.0 cents/kilowatt-hour, the energy needed for 1.0 kW for 1 hour.

P10.2. Your shower stall has two taps; the cold tap supplies water at 20.0°C, and the hot tap supplies water at 70.0°C. How would you adjust the relative flow rate hot/cold so that the mixture would be at the more comfortable temperature of 30.0°C?

P10.3. Calculate the rate of heat loss through a glass window in a house. Take the room temperature as 25°C, the outside temperature as −5.0°C, and the window area as 1.0 m^2 with a thickness of 3.0 mm. For the thermal conductivity of glass, k_T, take 0.040 J/(s-m°C).

P10.4. A medium sized candy bar has food value of about 200.0 Calories. You eat it in preparation for climbing Mt. Rainier. If the food value were completely converted into mechanical energy, what altitude gain would it take you to on the mountain?

P10.5. Rich ice cream has a food value of about 300.0 Calories per cup. What is the net energy value of one cup (mass ≈ 0.25 kg) of ice cream, which has a temperature of −10.0°C? It must be liquefied and warmed to body temperature, 98.6°F. Assume that the latent heat and the heat capacity of ice cream, when frozen and melted, are 80.0% of the value of ice (0.49 cal/g-°C) and water, respectively.

P10.6. An ice cube of mass 100.0 g and initial temperature −10.0°C is dropped into 200.0 g of water at an initial temperature of 40.0°C. Take the specific heat of ice as 0.49 cal/g-°C. When the system comes to equilibrium, is there any ice left? If not, what is the temperature at equilibrium? Assume that the system is thermally isolated from its environment.

P10.7. A certain Carnot engine has the following characteristics: $T_H =$ 900.0°C, $T_c =$ 150.0°C, $\Delta Q_H =$ 500.0 cal.

(a) How much heat is ejected into the cold reservoir?
(b) How much work is performed per cycle?
(c) What is the engine's efficiency?

P10.8. Estimate the temperature difference between the water at the top and bottom of Niagara Falls, due to the height change of 52 m. Assume that the mass of water is 1.0 kg/s.

P10.9. Estimate the atmospheric pressure at the top of Mt. Rainier, $h \approx$ 14,400 ft. Above sea level, from the following: the normal barometric pressure P at sea level is 760 mm Hg (mercury). The pressure at an altitude $h = 7000$ ft is 570 mm Hg. The variation of pressure with altitude is exponential, P (mm of Hg) $\propto e^{-kh}$.

P10.10. A 520 g piece of granite at 0.0°C, specific heat 0.19 cal/g-°C is immersed in 2.2 kg of water at 20.0°C.
(a) What is the equilibrium temperature?
(b) How much heat was given up to the granite?
(c) What is change of internal energy of the water?

Answers to odd-numbered questions

Q1. Mass, weight, volume, and energy are extensive quantities.

Q3. Yes, the regular change allows it.

Q5. The match needs convection to bring fresh air and oxygen; convection requires gravity.

Q7. Energy conservation is used to denote the use of less energy. The conservation of energy is a law of physics.

Q9. The thermal conductivity of metals is much greater than that of rubber. Thus, heat is conducted away from your hands much more rapidly by the handlebar.

Q11. Yes, you may change its phase.

Answers to odd-numbered problems

P1. 50 liters $\times$ 20° $\times$ 4.18 kJ/liter = 4.18 $\times 10^3$ kJ. 1 kW-hr = 3.6 $\times 10^3$ J. Thus, the energy of 1.16 kW-hr costs 5.8 cents.

P3. 400 W.

P5. $300 - 22.7 = 276$Cal.

P7. (a) $\Delta Q_c = 180.3$ cal; (b) $\Delta W = 320.3$ cal; $e = 0.639$.

P9. $P/P_o = 570/760 = e^{-k(7000)}$. Take the natural log of both sides: $7000 \ k = \ln(760/570) = 0.288$; $k = 4.11 \times 10^{-5}$ ft^{-1}. Then $P = \exp(-0.592)P_o = 0.553P_o = 421$ mm of Hg.

Chapter 11

States of Matter: Solids, Liquids and Gases

All of the chemical elements and many compounds have solid, liquid, and gas phases. This chapter describe them on a macro- and microscopic level. In the solid phase the molecules often form a *crystal lattice*, where they have regular positions, about which they can vibrate. In the liquid phase, the molecules are randomly packed together, with a little space in between, so that they can diffuse from place to place; the liquid can thus be stirred. At higher temperatures, the liquids vaporize and form gases.

11.1   Gases

A dilute gas is the simplest of all the phases. The atoms or molecules are, on average, far apart, with vacuum in between. They move in straight lines, occasionally collide, and bounce away (remember the excerpt from Lucretius' poem in Chapter 1). The steady rain of molecular impacts on the walls of the container that holds the gas exerts an average force on the walls that is proportional to the area.

In a gas or a liquid, *fluid*, it is not the force that matters as much as the *pressure*, which is the force per unit area perpendicular to that area. Think of yourself on snow. You are likely to sink into the snow if you do not have on some snowshoes. The snowshoes distribute your weight over a larger area and thus you exert less *pressure*. Pressure is

$$P = F_\perp/A \,, \tag{11.1}$$

where $F_\perp$ is the force perpendicular to the area. Pressure is measured in N/m^2 and $1 \ N/m^2 \equiv 1$ Pascal, abbreviated 1 Pa. The unit of pressure is

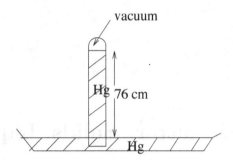

Fig. 11.1　A mercury barometer.

named after Blaise Pascal (1632–1666), who carried out studies of fluids. (He also invented a calculating machine.)

In the atmosphere, the pressure arises from the weight of the atmosphere above the point where you are measuring it. The pressure on the top of a mountain, e.g., Mt. Rainier, is thus lower than at sea level.

Pressure is measured by a *barometer*. To measure air pressure, a tube, closed at one end, is filled with a liquid and inverted in an open container of the same liquid, e.g. water. The air pressure on the open container can support a column of the liquid only up to a certain height. In the case of water, this height is about 32 feet or 10 m. Indeed, Galileo found that air pressure could not lift water any higher. But a water barometer is not very easy to carry around. A Galileo disciple, Torricelli, realized that mercury (Hg), being about 13 times denser than water would only rise 1/13th as high, 76 cm to be more exact, and this makes a reasonable barometer.

You may be unaware of atmospheric pressure because you are in equilibrium with it. You can demonstrate the effect of this pressure by evacuating the air from a closed aluminum container. It will collapse. In the middle ages, von Guericke evacuated two closely fitting and joined hemispheres. He then bet that two teams of (eight) horses would be unable to pull the hemispheres apart and won his bet!

11.1.1　*Perfect gases*

At this point, we know enough to be able derive the *perfect gas law*, and that temperature is a measure of the kinetic energy of the gas molecules. Furthermore, this derivation will give you an idea how physicists go about solving problems.

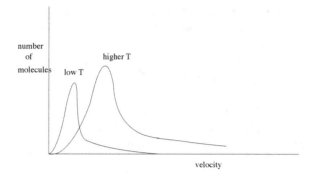

Fig. 11.2 Distribution of speeds of a gas at two temperatures.

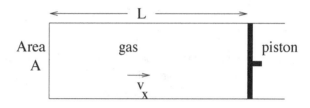

Fig. 11.3 A cylindrical container with gas.

We consider a "perfect" gas, where we make the following assumptions: (1) The molecules that make up the gas are so small that we can neglect their size. (2) The molecules have random motion with varying speeds. (3) The molecules exert no force on each other and you can neglect their collisions. (4) Collisions of the molecules with the walls, responsible for the pressure on the container, are elastic.

You cannot predict the motion of each and every molecule, but you can discuss the average properties of the motion. What is the average speed distribution? Maxwell studied this and found that the *root mean square* velocity $\sqrt{\bar{v}^2} \sim 400$ m/s. As the temperature of the gas increases the distribution of speeds, as shown in Fig. 11.2, has a peak which shifts to higher speeds and becomes broader.

Consider a gas in a horizontal (to avoid gravity effects) cylindrical container with a piston, as shown in Fig. 11.3. The cylinder has a length L and a cross sectional area A. Consider, further, a molecule with speed v_x moving in the $+x$ direction. When it hits the piston, it rebounds elastically with the same speed, but it is now moving in the $-x$ direction. The change in momentum of the molecule is thus $-2mv_x$. The force exerted on the piston by the collision is

$$\vec{F} = F_x = \frac{\Delta p_x}{\Delta t} = \frac{2mv_x}{\Delta t} \, . \tag{11.2}$$

Henceforth we call $\vec{F}$ simply F. Since the particle moves freely in the $\pm x$ direction until it hits the wall or piston, the time to make a second collision with the piston is $2L/v_x$. Thus the average force is

$$\langle F \rangle = \langle F_x \rangle = \frac{2mv_x^2}{2L} = \frac{mv_x^2}{L} \, . \tag{11.3}$$

The pressure on the piston is the force per unit area, or

$$P = \frac{\langle F \rangle}{A} = \frac{mv_x^2}{AL} = \frac{mv_x^2}{V} \, , \tag{11.4}$$

where V is the volume of the container.

There are, of course, a huge number of molecules in the container. They move in any direction. The average velocity is thus zero, since they are just as likely to move in the $-x$ direction as the $+x$ direction. However, this also means that

$$\bar{v}_x^2 = \bar{v}_y^2 = \bar{v}_z^2 = \frac{1}{3}\bar{v}^2 \, . \tag{11.5}$$

We thus find that the pressure due to all molecules is

$$P = \frac{Nm\bar{v}^2}{3V} = \frac{2}{3}\frac{N}{V}\bar{K} \, , \tag{11.6}$$

where N is the number of molecules and $\bar{K}$ is the average kinetic energy of a molecule. In summary, we obtain the *ideal gas law*

$$PV = \frac{2}{3}N\bar{K} = Nk_BT \, , \tag{11.7}$$

$$\bar{K}/\text{molecule} = \frac{1}{2}m\bar{v}^2 = \frac{3}{2}k_BT \, , \tag{11.8}$$

where k_B is the Boltzmann constant $= 1.38 \times 10^{-23}$ J/K. Equation (11.8) shows that the temperature of a gas is a measure of its kinetic energy, and Eq. (11.7) shows that the pressure times the volume is determined by the temperature of the gas, or its kinetic energy. With the assumptions we have made, the kinetic energy is equal to the internal energy because the potential energy is negligible.

The law is remarkably general. It does not matter what kind of molecule or atom it is, its mass or complexity, just how many there are. But it needs to be dilute enough, so that the size and the range of attraction of the molecules are a very small fraction of the average distance between them. All gases become ideal when they are very dilute. A small molecule, such as O_2, has a diameter of a few tenths of a nanometer. Notice that the product PV in Eq. (11.7) is a constant if the temperature is fixed; that is, the pressure and volume vary inversely with each other. Robert Boyle discovered the relation; it is called *Boyle's law*.

The number of molecules N in the Ideal Gas Law is sometimes expressed in terms of the number of n of its *moles*. One mole of a substance ... any substance ... consists of the same number of molecules, "Avogadro's number" $N_o = 6.02 \times 10^{23}$. A determination of N_o was the subject of one of Einstein's earliest papers. An alternative of the Ideal Gas Law expresses N in terms of N_o

$$PV = nRT, \tag{11.9}$$

where n is the number of moles, and $R = N_o k_B = 8.314$ J/mole-K is the *gas constant*. $N_0 m = M$ is the *gram-molecular weight* ("weight" is improperly used; it should be mass, but convention has it otherwise) of the substance in grams, where m is the "molecular weight" of one molecule. For example, the molecular weight of the oxygen molecule O_2 is 32; then one mole of molecular oxygen has a mass of 32 grams. At "standard conditions", $T = 20°C$ and $P = 1$ atm, the volume of one mole of an ideal gas is 22.4 liters.

Illustrative Problem: A perfect gas is in a 1 liter bottle at 20°C and atmospheric pressure.
(a) How many molecules are there in the gas?
(b) What is the total kinetic energy of the gas?

Answer: 1 mole takes up 22.4 liters and there are 6.02×10^{23} molecules. For one liter, there are $6.02 \times 10^{23}/22.4 = 2.69 \times 10^{22}$ molecules.
(b) The total K is $(3/2)k_B T = 1.5 \times 1.38 \times 10^{-23}$ J/K $\times 2.69 \times 295.2$ K $\times 10^{22} = 164$ J.

If a gas expands at constant pressure, P, the work done by the gas is $\Delta W = P\Delta V$. This comes from $\Delta W = F\Delta L = (F/A)(A\Delta L) = P\Delta V$.

11.2   Liquid State

As heat is extracted from a gas the average energy decreases; the molecules slow down, and the attractive forces cause them to spend greater fractions of time in each other's vicinity. The molecules begin to form temporary clusters. There are clusters of two, three, and more molecules; as the temperature drops, the clusters live for longer times, and grow in increasing size and number. At some point the clusters grow enormously into droplets of liquid, and if the pressure is maintained, the droplets grow and coalesce: the gas is *condensing; liquefying.* In the liquid the molecules are a dense, disorderly collection; their neighbors are close, but leave enough room so that small groups of molecules can swirl around, and the liquid can be stirred.

Several molecular gases are readily liquefied by cooling with ice or ice-salt mixtures, at temperatures down to about $0°F$ (for *Fahrenheit*). (In fact, the Fahrenheit scale set $0°$ as the lowest temperature that could be obtained by ice-salt mixtures). By the middle of the 19th century there remained several "permanent gases"; most of the components of air (oxygen O_2, nitrogen N_2, argon Ar and neon Ne), as well as hydrogen H_2 and helium He. James Dewar developed a process for liquefying air in commercial quantities. With liquid air as a primary cooling agent, all of the gases but He were liquefied. Dewar also invented the insulating "dewar flask", named after him. He was knighted for his achievements. The following poem plays on his name for a rhyme:

> *My name is James Dewar,*
> *and I'm smarter than you are!*
> *I showed lads and lasses*
> *I could liquefy gases!*

As an overall principle, cooling a material reduces its thermal disorder. This allows the subtle and fundamental properties to become prominent. For example, some materials are very weakly magnetic at ordinary temperatures, and become more strongly magnetic as the temperature falls, because thermal agitation is less disruptive to their orientations. More spectacularly, there are the *macroscopic quantum systems*; superconductors and superfluids, where quantum mechanics plays a role in large scale properties.

Finally, Helium was liquefied by a team led by H. K. Onnes at Leiden University in 1908. The normal boiling point (nbp) (The nbp of a substance is the temperature of condensation of the gas or vapor at normal atmospheric pressure, $\approx 10^5$ Pa. It is the temperature at which the liquid turns into a gas or vapor) of He is 4.2 K; He^4 is the dominant isotope; (see Chapter 23) the lighter isotope He^3 is present in very small concentration. Its nbp is ≈ 3.3 K.

When helium was liquefied it opened up an entirely new range of temperatures, much colder than its normal boiling point. It is cooled by pumping its vapor away; by *forced evaporation*. This is the same process that chills you when you come out of the shower or a swimming pool; the latent heat of evaporation drains heat away from the remaining liquid and its container; you. As long as the liquid He is in a well-insulated vessel, such as a dewar (a double walled flask, with vacuum between the walls, which James Dewar invented), and shielded from infrared radiation, the remaining liquid is readily cooled to about 1 K. One of the first things that Kammerlingh Onnes did with liquid helium was to cool a sample of mercury and measure its resistance. He was testing a hypothesis of Lord Kelvin, that at a sufficiently low temperature, metals would become insulators. What Onnes found instead, was absolutely the opposite: At a temperature ≈ 4 K the resistance dropped to an immeasurably low value; the mercury became a *superconductor*. At first, Onnes didn't believe what he saw; he and his team repeated the experiment until they were convinced. To demonstrate that the resistance of superconducting lead was truly zero, a persistent current was induced in a ring; it continued flowing for a year, until the ring was warmed above its transition temperature. Soon after finding the effect in mercury, they found it also in tin and lead. Superconductivity could not be understood until John Bardeen, Leon Cooper and J. Robert Schrieffer gave a microscopic theory in 1957, for which they were awarded a Nobel prize in Physics. In 1986 Bednorz and Mueller discovered the first high temperature superconductor, compounds that have their transitions near or above the nbp of liquid nitrogen. This made them commercially practical for many applications, such as levitating trains. Further discoveries followed, and continue today. There are many known superconductors, both elemental and compound. There are other superfluids and macroscopic quantum systems, subjects that are outside of our syllabus.

11.2.1 *Hydrostatics*

In a uniform fluid the mass m of every small volume v is proportional to the volume; the *density* $\rho = m/v$ is the same throughout the total volume

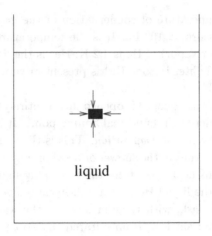

Fig. 11.4   Pressure on a small box in a liquid.

V. The total mass M is therefore

$$M = \rho V.$$

The density of water at $4°C = 1$ g/cm$^3 = 1.00 \times 10^3$ kg/m^3. The density of ocean water is higher due to its salt content.

In a fluid, the pressure acts *out of the fluid* on any object, for instance on the walls of the container holding the fluid. In Fig. 11.4, we show the pressure on a small box placed in the liquid.

Any increase in the pressure of a fluid is transmitted uniformly throughout the liquid without diminution. This was first found by Blaise Pascal in France and is called *Pascal's principle.*

Hydraulic jacks make use of Pascal's principle. A small force applied to a small area can lift a large weight (e.g. a car or truck) supported by a large area. See Fig. 11.5. Although there is a *mechanical advantage* in a hydraulic jack, you do not gain any energy. The work done at the little area a is fL, where L is the height you have to push through. The work done at the large area, A, is $F\ell$, where ℓ is the height the large piston rises as the little piston is pushed down.

$$\Delta W = fL = F\ell, \tag{11.10}$$

and

$$F/f = A/a. \tag{11.11}$$

This is called the mechanical advantage.

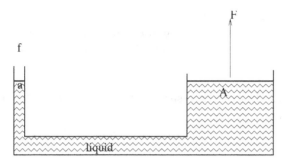

Fig. 11.5   A hydraulic jack.

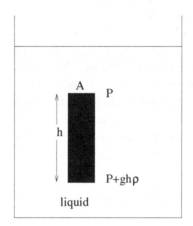

Fig. 11.6   Increase in pressure with depth due to weight of a column of liquid.

In a fluid, pressure increases with depth due to the weight of the liquid above, as it does in air. The weight of a column of liquid of height h and area A is $\rho A h g$, where ρ is the density of the liquid. The pressure therefore increases with depth h as $\rho g h$. This is illustrated in Fig. 11.6.

At a depth h in a fluid of density ρ, the increase in pressure relative to the surface is

$$\Delta P = \rho g h \,, \qquad (11.12)$$

where h is the depth. Since water is nearly incompressible its density is independent of depth. The pressure under water increases at the rate of about 1 atmosphere for every 10 meters of depth.

Illustrative Example: Water fills a container to height of 22 cm. Find (a) the excess pressure over air pressure at the bottom of the vessel and (b) the speed at which water would escape from a hole in the bottom of the container.

Answer: (a) The excess pressure is $\rho g h = 10^3 \text{ kg/m}^3 \times 9.8 \text{ m/s}^2 \times 0.22 \text{ m} = 2.2 \times 10^3$ Pa.

(b) K.E. $= \frac{1}{2}mv^2 = mgh$, or $v = \sqrt{2gh} = \sqrt{2 \times 9.8 \text{ m/s}^2 \times 0.22 \text{ m}} = 2.1$ m/s.

Jules Verne's science fiction classic *20,000 Leagues Under the Sea* tells of a submarine that travels through all the World's oceans (20,000 leagues refers to the submarine's horizontal travels; the maximum depth mentioned in the book is much less: 4 leagues. That is still considerable: $\approx$ 23 km!!) at such a depth the pressure is greater than 2000 atm, a greater pressure than any real submarine can withstand. Yet there are sea creatures, such as sperm whales, that can dive to great depths (although not quite so deep). How can this be? The pressure exists throughout their bodies; while they dive or rise the pressure is continually equalized throughout their bodies. A submarine, with its rigid shell and internal air environment, cannot so respond.

11.3   Buoyancy

Why does a solid piece of wood float and the same size metal block sink? When an object is immersed in a fluid it is *buoyed up* by the weight of the volume of fluid it displaces.

The increase in pressure with depth of fluid also provides an upward (buoyant) force. Since the pressure is the same in all directions, the upward force on the bottom of an object immersed in the fluid (as in Fig. 11.6) is just $\rho g h A = \rho V g$, where V is the volume of the cylinder. This upward force is called *buoyant* force. The *buoyant force* exerted on a partially or entirely submerged object is just equal to the *weight of the volume of fluid displaced*. This is known as *Archimedes' principle*.

Note that you have to do work to submerge a piece of cork. Why? There is an upward buoyant force that you must work against to submerge it.

As the story goes, King Midas commissioned an artisan to fashion a gold crown, and provided the metal to make it. When the crown was delivered, Midas suspected that the artisan, in order to make an illegal profit, might have withheld some gold and alloyed the remainder with a cheaper metal. (Gold can be alloyed with appreciable amounts of copper or silver, without a great change of color). Archimedes was tasked with the problem of testing the crown. He knew that alloying would reduce the density; he could weigh the crown, but the problem was how to determine the volume of the intricate shape. As the legend goes, he hit upon the solution ... *Archimedes Law* ... while in his bath. And he was so excited that he ran out into the street, shouting "Eureka"! ("I found it!") It was to compare the weights of the crown in air and immersed in water. Thus, the change in weight would be due to the buoyant force: the weight of the displaced water, and therefore directly related to its volume.

> *What is its weight in the water?* That is the nub
> Of the question Archimedes solved in his tub.
> *Buoyed up by the weight of the water displaced,*
> Solved the king's problem, his goldsmith disgraced.

An object will sink until its weight is balanced by the buoyant force, i.e. until the weight of the volume of fluid displaced is equal to the weight of the object. For a uniform object of density ρ, the fraction of the object above a given liquid depends on the density of the object, ρ_o, relative to the density of the fluid, ρ_f. If $\rho_o < \rho_f$, a portion of the object will be above the fluid. The fraction under water is such that the weight of the fluid displaced is equal to the weight of the object, $\rho_f x = \rho_o h$, where h is the height of the object. If you multiply by the area of the object (see Fig. 11.7), then $\rho_f x A g = W_f$ is displaced and $\rho_o h A g = W_o$, where x is the distance the object is submerged and h is the entire height of the object of cross sectional area A.

Illustrative Problem: If you want to salvage a 2000 kg cube of 1 m sides at 2.2 m depth of water, what is the minimum force (no acceleration) needed to bring it to the surface?

Answer: The buoyant force is the weight of water displaced $= 1$ m$^3 \times 10^3$ kg/m$^3 \times 9.8$ m/s$^2 = 9.8 \times 10^3$ N. The cube's weight in water is 2000 kg $\times$ 9.8 m/s$^2 - 9.8 \times 10^3 = 9.8 \times 10^3$ N and this is the minimum force needed.

The buoyant force can be large enough to float large and massive solid objects. The object sinks down into the water until the weight of the

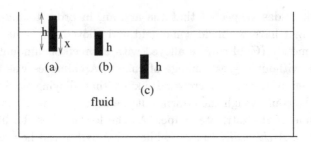

Fig. 11.7 An object of height h, (a) with $\rho_o < \rho_f$, (b) with $\rho_o = \rho_f$ and (c) with $\rho_o > \rho_f$.

displaced volume is equal to the weight of the object. And so there are ships of steel, and boats of concrete, perfectly seaworthy as long as there is enough "empty" volume to displace enough water, allowing the vessel to stay well above splashes and waves. A more complicated but equally important question is the vessel's *stability*, how large an angle of tilt it can sustain and tend back to its upright position. But, we will not examine stability here.

Hot air balloons or He filled balloons work on the same principle. When air is heated, its density is reduced and He is lighter than air. The balloon and its contents float and rise until their weight is equal to the weight of air displaced by the volume of the balloon.

11.4 Flowing Liquids

What causes water in a river to flow? It is a difference of pressure, ΔP or gravity. If you have watched the water in a river, you know that the speed in the middle is larger than close to the shore, and that the water speeds up if the channel narrows. Why? The answer for the first observation is the friction of the water at the sides. There is also friction between layers of the fluid, called *viscosity*.

If there is no water lost or gained (e.g., by a stream entering the river), then we say that the flow is *continuous*, and the same amount of water leaves the river at the ocean as there was at the beginning. This neglects the loss of water by evaporation. The rate of flow is the volume, V, of water that passes a given location per unit time, $V/t = AL/t = Av$, where A is the cross sectional area of the river and L is the given length of the river that passes the location in the time t. Here v is the speed of the river at

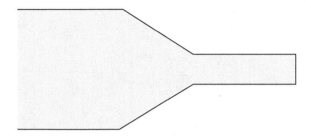

Fig. 11.8 In a pipe, the velocity of a fluid depends on the cross sectional area of the pipe.

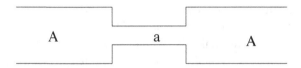

Fig. 11.9 A pipe that narrows and then gets larger again.

the given location. The rate Av is constant, since no water is lost. Thus if the area increases, the velocity of the river is lower so that Av is constant. See Figs. 11.8 and 11.9, where $Av = av'$, with v' larger than v.

Illustrative Example: A rowboat floats by at 2 mph in a river that is 8 ft wide. If the river narrows to 2 ft, what will be the speed of the boat?

Answer: Since vA is constant, the speed will be (8/2) 2 mph = 8 mph

There are other features of flow — it can be smooth or turbulent. This feature depends on the viscosity of the fluid. The higher the viscosity, the harder it is to produce turbulent flow.

11.4.1 *Bernoulli's principle*

It is commonly believed that when the velocity of a liquid increases, as in the middle section of Fig. 11.9, that the pressure increases. This is not correct. The conservation of energy guarantees the opposite! It leads to *Bernoulli's principle*, published by Daniel Bernoulli in 1738,

$$P + \frac{1}{2}\rho v^2 = \text{constant}. \qquad (11.13)$$

Fig. 11.10   The causes of a lift of an airplane wing.

Why is this conservation of energy? For a fluid we have to use quantities per unit volume. The K.E. per unit volume is $\frac{1}{2}mv^2/V = \frac{1}{2}\rho v^2$. The work done on the fluid is force × distance$/V = (F/A)(L/L) = P$. Thus the work done + K.E. = constant. If the height of the fluid changes, we have to add the change in gravitational P.E./volume $= \rho g \Delta h$.

There are numerous applications of Bernoulli's principle. You may have seen a ping-pong or other light ball being held up by a vertical blower in front of a store. Why is the ball stable? If it moves to the side, the velocity of the air decreases and the pressure increases, forcing the ball back to the center. You can try this at home with a hair dryer, but you need to center the ball and not get burned in the process!

Likewise, if you blow across the top of a light piece of paper, it will rise because the increased velocity over the top lowers the pressure, forcing the paper to rise. The lift of an airplane partially works on the same principle. The shape of the wing is such that the flow of air over the top is more rapid than over the bottom. This causes a lift due to the increased pressure on the bottom of the wing. However, there is another effect that comes in: this is a reaction force. The stream of air is made to deflect downwards. By Newton's laws there is then a reaction force that tends to lift the wing. An illustration is shown in Fig. 11.10.

Bernoulli's principle can also explain the curve of a baseball expertly thrown by a pitcher, as well as similar phenomena.

Questions

Q11.1. Why does blowing on soup cool it?

Q11.2. Why is there a vacuum on the top part of a barometer? (see Fig. 11.1)

Q11.3. Why do you feel chilled coming out of a shower?

Q11.4. Do molecules in a relatively dense gas travel in a straight line? Explain.

Q11.5. Why does chalk sometimes make scratching sounds on a blackboard?

Q11.6. In view of the high speed of molecules at room temperature, why does it take a millisecond or longer to smell a flower from far away?

Q11.7. Two playing cards are fairly close together. If you blow between them, will they separate or approach? Why?

Q11.8. If you fill a drinking glass with a carbonated beverage and toss in some raisins, then after the foam has settled, the raisins will sink to the bottom, but soon rise again to the surface. After a few seconds, they will dive again. Explain. Try it!

Q11.9. In a gas of hot air, is the speed of oxygen and nitrogen molecules the same? If not, which goes faster and why?

Q11.10. What is it that limits water from being brought up from a depth of more than 10 m without mechanical help?

Q11.11. Give at least four properties of pressure in a fluid. An example would be that pressure is transmitted throughout the fluid without loss.

Q11.12. A row boat in a pond contains an anchor. The anchor is dropped overboard. At first, the anchor rope is not long enough for the anchor to reach bottom, but then the rope is lengthened and the anchor rests on the bottom.
(a) At first, will the boat rise higher, sink lower, or stay the same in the water. Explain.
(b) Will the water in the pond fall, rise, or remain at the same level when the anchor reaches the bottom? Explain.

Q11.13. Is the pressure on the bottom of the container in Fig. 11.5 the same everywhere? Explain.

Q11.14. A gas has its temperature raised. Is work done on or by the gas? Explain.

Problems

P11.1. At what depth of water will the pressure be twice atmospheric pressure?

P11.2. Mariana's trench is 1.13 km deep under water. What is the pressure due to the water above it?

P11.3. An iceberg has a density of 0.86 that of sea water. How much (what fraction) of the iceberg will be above water?

P11.4. The hydraulic lift in a service station has a large cylinder of 60.0 cm diameter; the actuator piston has a diameter of 5.0 cm.
(a) What force must be applied to the small piston to raise a 2100 kg truck?
(b) How much must the actuator cylinder go down to raise the truck 12 cm?

P11.5. A cargo barge has the shape of a large box, with dimensions 80.0 m long, 30.0 m wide, and 15 m deep. It is constructed of 1.0 cm thick steel plate.
(a) If the density of the steel is 7.0 g/cm^3, what is the mass of the barge?
(b) How far below the surface will the bottom of the barge be in water?

P11.6. An object "weighs" (its mass is) 5.0 kg in air and 4.0 kg when immersed in water. What is the volume of the object?

P11.7. A river is 40.0 m wide and 2.2 m deep. Water flows with a speed of 4.0 m/s. The river goes through a gorge of width 4.0 m at a speed of 6.0 m/s. What is the depth of the water in the gorge?

P11.8. Water fills a cylinder of diameter d to a height h. What will be the speed of the escaping water if a small hole occurs at a height $h/2$?

P11.9. What is the work done by a gas in doubling its volume at constant pressure? Does the temperature of the gas increase, decrease, or remain constant in the process? Explain.

P11.10. A river goes through a narrow channel of the same depth, but 1/3 of the width.
(a) what is change in velocity, i.e., v(channel)/v(normal)?
(b) Find the ratio of the pressure in the channel to the normal pressure.

P11.11. How many molecules of oxygen are in 1.0 m^3 of air at 20°C and normal atmospheric pressure. Assume that the concentration of oxygen in air is 20%?

P11.12. What is the minimum volume of a hydrogen balloon that can lift a 500.0 kg gondola and its own skin of 300.0 kg at sea level? The molecular weight of hydrogen is 2.0 g/mole and the average molecular weight of air molecules is 29 g/mole.

P11.13. (More difficult) How hot must the air be in a 100.0 m^3 balloon to provide 100.0 N of lift if the outside air is at 20.0°C. Assume that the density of air is inversely proportional to the temperature, and that at 1 atmosphere pressure and 20°C it is 1.30 kg/m^3.

P11.14. What is the average K.E. of an oxygen molecule at a temperature of 220°C? What is it at 22 K? What is its average speed. (The mass of an oxygen molecule is 32/N_o grams).

Answers to odd-numbered questions

Q1. Because the blowing removes the vapor emanating from the soup, which allows more evaporation.

Q3. Because of the evaporation.

Q5. The friction may cause the chalk to move in a "stick-slip" mode; that causes high frequency vibration.

Q7. They will approach because the higher speed between the cards means less pressure.

Q9. The nitrogen will move faster, since the kinetic energies are the same, but the mass of nitrogen is less than that of oxygen.

Q11. Pressure is the same at the same depth; pressure is always perpendicular to any surface and out of the fluid; pressure is the same in all directions; the difference in pressure with depth below surface is $h\rho g$.

Q13. It is the same, because the pressure is transmitted unhindered.

Answers to odd-numbered problems

P1. At 10 m.

P3. 14%.

P5. (a) There is 81 m^3 of steel with a density of 7000 kg/m^3. Thus the mass is 5.67 × 10^5 kg.

(b) The barge will sink until it displaces its weight of water. Thus $2400h\rho = 5.67 \times 10^5$ and $h = 0.24$ m or 24 cm.

P7. $h = 4 \times 40 \times 2.2/(6 \times 4) = 15$ m.

P9. Work $= P\Delta V$. Since V increases and P is constant, the temperature increases.

P11. 1 m $= 10^3$ liters. The number of molecules is $10^3 N_o/22.4 = 2.69 \times 10^{25}$. Oxygen is 20% thus has 5.38×10^{24} molecules.

P13. The weight is $\rho V g = 1.30$ kg/$m^3 \times 10^2$ $m^3 \times 9.8$ m/$s^2 = 1.27 \times 10^3$ N; also $\rho_T/\rho_{20} = (293.2/T)$. We have $1.3 \times 10^2 \times 9.8 \times (1 - 293.2/T) = 100$ N. Thus $T = 295.6$ K or 22.4°C.

PART 2
Electricity and Magnetism

Chapter 12

Static Electricity

12.1 Amber, Amusements, and the Beginnings of Serious Study

Amber is fossilized tree sap, a gold-colored, transparent solid. It was prized in antiquity as "the golden gemstone of the forests", and it was carved into ornaments, amulets, and figurines. It was also known as "the golden attractor" because leaves, feathers and thread were attracted to it after it had been rubbed. When a spinner's distaff (a rod that held wool or flax, to be drawn and twisted into thread) was made of amber, it held the thread so tightly that it was called "the clutcher". The Greek word for amber is "elektrum". For over 1000 years it was the sole substance that was known to have the "amber effect".

Other materials were gradually discovered which were amber-like, in other words, *electric*. In the 16th century diamond was found to be electric, then other gems: sulfur, sealing wax and glass.

The Ancients were not at a loss to explain the amber effect, but their understanding was on a different level from the modern meaning of the word. Plato wrote that one should believe the eye of the mind rather than the eye of the body, because the true order and beauty of the universe was obscured from our sight by the dust and detritus of the everyday world. Earlier Greeks had trusted observation more, but by the time of Aristotle, the philosopher-naturalists had abandoned the reliance on the physical senses. Truth sprang from the inner sight, and its revelations could be extraordinarily detailed. Plato wrote, in "The Timaeus":

> *Furthermore, as regards all flowings of waters, and fallings of thunderbolts, and the marvels concerning the attraction of amber and of the Heraclean*

stone (lodestone, the magnetic ore of iron) not one of these possesses any real power of attraction; but the fact that there is no void, and that these bodies propel themselves round one another, and that according as they separate or unite they all exchange places and proceed severally each to its own region — it is by means of these complex and reciprocal processes that such marvels are wrought.

This is an example of nonsensical "explanation" of natural science and medicine before the Renaissance. If a date can be put on the break from the classical natural philosophy, it is 1600, with the publication of William Gilbert's *De Magnete*. It was an enormously influential book, not only for his discoveries, but also for pioneering the scientific method. Little is known of Gilbert's life and career beyond the following essentials, owing to the loss of records in two events; the destruction of portions of his town of Colchester during the Second Civil War, and the Great Fire of London. Gilbert was born in 1544 in Colchester, the son of a well-to-do Recorder of that town. At age 14, he matriculated as a student at St. Johns College of Cambridge University, received a B.A. degree in two years and his masters degree four years after that. He remained in college for five years more, that is until 1569, becoming a medical doctor and senior fellow. He settled in London and began a medical practice in 1573. Toward the end of a distinguished career he served as President of the Royal College of Physicians, and in 1600 was appointed one of the physicians to Queen Elizabeth. He died in 1603, apparently of the plague, less than a year after the Queen's death.

We know even less of Gilbert's life during the period when he carried out most of his experiments, between leaving college and the beginning of his London practice. There is some evidence that he traveled on a grand tour of the Continent, during which he spent four years in Italy. The publication of *De Magnete* came many years later. It expressed a new attitude toward worldly matters and experience, one of the clearest statements of the Renaissance's break with classical scholarship. The Preface laid out the new philosophy.

Clearer proofs, in the discovery of secrets, and in the investigation of the hidden causes of things, being afforded by trustworthy experiments, and by demonstrated arguments, than by the probable guesses and opinions of the ordinary professors of philosophy.

Whoso desireth to make trial of these same experiments, let him handle the substances, not negligently and carelessly, but prudently, and deftly, and in the proper way; nor let him (when a thing doth not succeed) ignorantly denounce our discoveries: for nothing hath been set down in these books which hath not been many times performed and repeated amongst us.

This nature-knowledge is almost entirely new and unheard-of, save what few matters a very few writers have handed down concerning common magnetical powers. Wherefore we but seldom quote ancient Greek authors in our support, because neither by using Greek arguments nor Greek words can the truth be demonstrated. For our doctrine magnetical is at variance with most of their principles and dogmas.

Gilbert's book is mostly concerned with magnetism; only a small fraction deals with electricity. However its importance far outweighs the number of pages. It describes experiments that show the amber effect to be shared by a large number of substances. Before Gilbert only one other substance — jet (a hard coal) — was known to show the amber effect. Gilbert detected it in gems, glass, sulfur, rock salt and sealing wax; he lists more than twenty. Gilbert made his discoveries with a simple but sensitive instrument of his own invention, which he called a "versorium". He describes its construction and challenges his readers to repeat and verify his experiments. Inspired by the magnetic compass, it consisted of a rotating needle

... of any metal you like, three or four digits in length, resting rather lightly on its point of support after the manner of a magnetick needle.

His research with the versorium revealed the earliest essential features of the science of electricity. A summary of his discoveries counts over thirty distinct properties, of which the most significant are the following.

Many materials, which Gilbert termed electrics, in addition to amber and jet, can attract small objects after being rubbed. They include diamond, sapphire, amethyst, opal, beryl, crystal, glass, sulfur, rock salt and sealing wax.

*Electrics attract not only straws and chaff but **everything**: all metals, woods, leaves, stones, earths, water and oil.*

Amber heated without rubbing will not attract.

The attraction of an electric for a versorium is greater if they are closer together.

A piece of amber will distort a drop of water into a cone, by attraction of its various parts.

A piece of thin silk interposed between an electric and an object will decrease the effect but not entirely remove it.

An electric draws an object in a straight line toward it.

With confidence in his own methods and results, he attacked the conventional wisdom with stinging language. Gilbert was an exuberant critic.

> ... *utterly false and disgraceful tales of the writers; absurd, dubious and untrustworthy. Useless phantasms. They treat the thing with words alone, without finding any proofs from experiments, their very statements obscuring the thing in a greater fog.*

He permitted himself

> *to philosophize freely and with the same liberty which the Egyptians, Greeks and Latins formerly used in publishing their dogmas; whereof very many errors have been handed down in turn to later authors, and in which smatterers still persist, and wander as though in perpetual darkness. To those early forefathers of philosophy, Aristotle, Theophrastus, Ptolemy, Hippocrates, and Galen, let due honor be paid: for by them wisdom hath been diffused to posterity, but our age hath detected and brought to light very many facts which they, were they now alive, would gladly have accepted.*

Giordano Bruno was burned at the stake in the same year, for heresies no greater.

Gilbert's explanations, however, were not as successful as his observations. They were less fanciful than the theories he criticized, but they did not escape completely from the belief in "humors" and "forces of sympathy". The importance of *De Magnete* was much greater for its exposition of a new philosophy of inquiry than for its account of results. It set forth a rationale for the

> *Scientific Method; a firm foundation on reproducible experimental observations and tests of the theories constructed to explain them.*

Gilbert experimented with a lodestone model of the Earth, with which he attempted to explain gravity as a magnetic attraction. He strongly supported the Copernican heliocentric theory that the Earth is a planet of the sun. Gilbert's book influenced Galileo, who repeated and extended some of his experiments on magnetism.

The poet Dryden said *Gilbert shall live till lodestones cease to draw.*

Gilberts achievements are all the more remarkable because they appeared in an England totally uninterested in physical science. There was great ferment certainly; the Elizabethan age knew a great increase in exploration, wealth and the growth of empire. Popular entertainments ranged from Shakespeare and the readings of poets to bear baiting. But intellectual life was frozen in its devotion to classical learning. Official statutes in Oxford and Cambridge declared that bachelors and masters of arts who did not adhere to the teachings of Aristotle were liable to a fine of five shillings for every point of divergence. Student life combined the seclusion of the monastery with the riotous dissipation of the tavern. Giordano Bruno, visiting Oxford and Cambridge, found the dons, despite their gorgeous robes and insignia, *as devoid of courtesy as cowherds.*

An understanding of the amber effect began with the discoveries of a few experimenters in the 18th century. But before then, electrical effects were "entertainments", where itinerant "experts" put on public shows of electric phenomena. A popular demonstration was to hire a poor street urchin, who allowed himself to be suspended by cords from a wooden frame. The demonstrator would charge a glass rod by rubbing on a cloth and then touch it to the boy. When this was done several times, he became sufficiently *electrified*, so that feathers, bits of paper and leaves were attracted to his fingers and toes, and sparks could be drawn from his nose.

The French king appointed a "Court Electrician", one Abbe Nollet, to demonstrate experiments at the royal court. Nollet had an even more dramatic demonstration. He had a number of soldiers line up, each holding a metal wire connecting himself to the next man. The first soldier grasped one electrode of a highly charged capacitor, and when the last in line touched the other electrode, a shock ran instantly through the line, and every soldier jumped at once "as if in formation".

Benjamin Franklin's interest in electricity began about 1774, (see Figs. 12.1 and 12.2), when he saw a street entertainment. A few years later Mr. Collinson, their book agent in London, sent him a glass rod and a magazine describing how it might be used to create electrical phenomena. Franklin was delighted; he gathered a few friends; they repeated the experiments and then began to study the effects seriously. By this time other materials had been discovered that could be electrified by rubbing, including glass, hard coal, sulfur, and several gemstones. It was believed that these materials could be divided into two classes: "resinous" and "vitreous". Members of the same class repelled each other, but attracted members of the other class. However, their actions were sometimes erratic. In Germany, Otto von Guericke invented a device that speeded up electrification; simply a sphere of sulfur mounted on a shaft that could be turned rapidly while a hand or a cloth was pressed against it. Guericke could carry the electrified sphere, which repelled a feather and floated it aloft while he walked about his laboratory. However, sometimes the sphere attracted the feather. Guericke concluded that it was up to the sphere to decide; "When it wants to attract, it attracts."

Franklin and his friends began their experiments in a spirit of playfulness, but soon discovered unexpected effects; their "play" turned into research. Franklin described their discoveries in a series of letters to Collinson, who happened to be a member of the Royal Society. Collinson read the letters to the members at their meetings, and the findings were so extensive and interesting that the collection was published as a pamphlet, that came to be known as *Franklin's textbook*. Some of the discoveries are listed below, in the words originally used to describe them.

Electrical phenomena seem to be due to a *single fluid*; rubbing can cause a deficit of the fluid, which leaves the object in what can be called *a negative state*, or an excess of the fluid, which causes *a positive state*.

The fluid is neither created or destroyed; it is simply transferred from place to place. The hand or cloth that rubbed the object, became positive if the object was made negative, and *vice versa,*

Like states repel, unlike states attract.

An electrified object has a *charge*. It can be *discharged* by *grounding*, i.e. touching with a metal object in contact with the ground. Electrons flow freely to and from the ground.

An electrified object can be discharged at a distance by a sharply-pointed and grounded metal rod.

Lightning is an electrical discharge.

Glass, when rubbed with silk, becomes positively charged. (This assignment of polarity was totally arbitrary ... it could just as well have been negative. But Franklin's choice is the basic reason why the electron's charge is negative).

12.2   Lightning is an Electrical Discharge

Lightning is one of Nature's most dramatic and frightening phenomena, e.g., Fig. 12.1. It has inspired myths and religious beliefs. The most important god of the Hittites was the weather god Teshub. He was similar to Zeus, and is often shown standing alone with a symbolic bolt of lightning. It was natural for the Hittites, who lived in a land (Anatolia, in present day

Fig. 12.1   Lightning discharge — lightning strikes near Oradea, Romania in 2005. Courtesy of Wikipedia.

Turkey) subject to frequent violent storms, to give precedence to a god of thunder and lightning.[a]

In the Middle Ages, all over Germany, the church bells were rung during thunderstorms, in the belief that it would protect them from harm. (They felt that the danger stemmed from the noise of thunder, rather than the lightning bolt). In some places, the sexton was paid to ring the bell as late as the middle of the 19th century. In the neighborhood of Constanz, all the bells in church steeples around the countryside were set ringing by volunteers, who believed that the sound would afford them complete protection from lightning, although some bell ringers were killed by lightning strokes in the very act of pulling the rope. There was a similar bell ringing practice in Paris, using the great bell of the Abbey of St. Germain. In old St. Paul's cathedral in London, there was a special endowment for *ringing the hallowed bell in great tempests and lightenings*. Similar rites were practiced in parts of Africa and China.[b]

---

[a]L. Cottrell, "The Penguin Book of Lost Worlds", v.2. Penguin 1966, 124–126.

[b]J. G. Fraser, "Folk-Lore in the Old Testament", Turdor Pub., N.Y., 1923.

Fig. 12.2   Portrait of Benjamin Franklin in 1767, by David Martin. Courtesy Wikimedia Commons.

Franklin proposed an experiment to test the hypothesis that lightning is a form of electricity. Many people had suggested a connection; Franklin devised a test. A tall sharply pointed metal rod would be raised above a "sentry box" where a man could stand on an insulated platform, and he could hold a wire from the rod by an insulated handle. When a storm passed overhead, perhaps some of the storm's electricity would be drawn off by the rod. Then the experimenter could bring the wire close to a grounded post inside the building, and that should cause a spark between them. Franklin wrote,

> *If any danger to the man be apprehended (tho' I think there would be none), let him stand upon the floor of his box, and now and then bring near to the rod the loop of a wire, that has one end fastened to the leads, he holding it by a wax-handle. So the sparks, if the rod is electrified, will strike from the rod to the wire and not affect him.*

In these words Franklin proposed the experiment in one of his letters to Collinson. The letter was included in the published collection. Collinson

Charles Augustin de
Coulomb

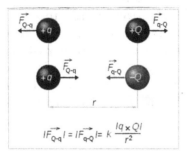

This picture shows how Coulomb's
Force act; similar charges pushing
against each other and opposite
charges attract each other

Fig. 12.3   Charles A. de Coulomb and an illustration of Coulomb's law. Courtesy of
Wikipedia.

sent a copy of the pamphlet to Bufon, a famous French naturalist, who
arranged for the printing of a French translation. And thus it caught the
attention of two French scientists, Dalibard and Delors, who decided to
perform Franklin's test of lightning's electric nature. A "sentry box", with
rod and wire, prepared to Franklin's specifications, was erected on a hill
at Marly-la-Ville (near the site of the present Charles de Gaulle airport).
Then they hired an old soldier to wait in the sentry box for an "electrified
cloud". On May 10, 1752, when a storm cloud passed overhead the soldier
brought the wire near the rod; there was a hissing sound, a great spark and
the smell "of burning sulfur". (We think rather than sulfur it was the smell
of ozone, the molecule O_3, which is produced in electrical discharges). The
old soldier shouted in alarm, which brought the village priest running with
a crowd of villagers. The experiment was repeated, and again there was
the hiss, the spark and the smell.

The priest wrote to Dalibard, "I am announcing to you ... what you have been waiting to hear; the experiment is finished ... I performed it six times, in the presence of several people." He would have continued, but the storm had died down. A few days later, Dalibard repeated the experiment. He wrote to France's Royal Academy of Sciences, "Mr. Franklin's idea has ceased to be a conjecture; here it has become a reality."

In the course of that summer, the experiment was repeated in England, Germany, Italy, Holland, Sweden and Russia. Franklin was famous all over Europe before the news reached him in Philadelphia.

Franklin suggested that buildings might be protected from lightning strokes by tall grounded rods with sharpened ends. His idea was disputed by a distinguished scientist, who thought rounded ends would be better. And by King George, who sought support for his decision from Sir John Pringle, the President of the Royal Society. However, Pringle went for pointed ends, and he defended his decision by saying that "he could not change a Law of Nature". The King replied, in that case Pringle ought to resign, so that the King might appoint someone who could! The dispute inspired a poem:

> *While you, great George, for safety hunt*
> *and sharp conductors change for blunt*
> *the Nation's out of joint.*
> *Franklin a wiser course pursues;*
> *and all your thunder, fearless views,*
> *by keeping to the point.*

To this day, lightning protection codes recommend Franklin lightning rods or some variant of them.

12.3   Coulomb Force

In the 1730's the French scientist Charles de Coulomb (Fig. 12.3) measured the strength of the electrical force. He applied the principle of the torsion balance that had been used by Cavendish to measure gravitational forces (Chapter 8). The *Coulomb force law* has the same form as gravitation. It is proportional to the product of the two charges, and inversely proportional to the square of the distance between them

$$F = k\frac{q_1 q_2}{r^2}, \qquad\qquad (12.1)$$

where the *Coulomb constant* $k = 9 \times 10^9$ Nm^2/C^2. The charge unit *Coulomb*, abbreviated C, is set by the force law. The interaction between two charges of 1.0 C each is enormously stronger than gravitation between two masses of 1.0 kg, owing to the great disparity between the constants G and k. The names that Franklin chose for the two types of charge, *positive* and *negative* are fortuitous; they automatically indicate, by the sign of the product $q_1 q_2$ whether the force is attractive (negative) or repulsive (positive).

Illustrative Example 12.1: Two charges, A, of 0.2 μC and B, of 1.2 μC are placed 2 m apart. What is the force that A exerts on B?

Answer: The force is one of repulsion of magnitude
$\frac{9 \times 10^9 \times 0.2 \times 10^{-6} \times 1.2 \times 10^{-6}}{4} = 5.4 \times 10^{-3}$ N.

12.4 Insulators and Conductors

Ordinary matter contains equal amounts of positive and negative charge; the matter is *electrically neutral* in its normal state.

Ordinary matter is composed of combinations of chemical elements. Each atom has a characteristic number Z of negative *electrons* surrounding a positively charged *nucleus*. Each electron has a charge $-e = -1.60 \times 10^{-19}$ C, and the nucleus has a balancing positive charge $+Ze$.

In insulating materials the electrons are tightly bound within the individual atoms or molecules. Electric fields may displace from their equilibrium positions, but as long as the fields are not too great, the electrons remain attached to their "home" atoms.

An atom that has lost one or more electrons is a positively charged *ion*. An atom that has gained one or more electrons is a negative ion.

In conductors there are mobile charges. In metals the mobile charges are electrons, but they are only a minor fraction of all the electrons, for most remain bound to their atoms. The "conduction electrons" move through a "bed" of positive ions. Liquid solutions of chemical salts, such as common table salt in water, are *ionic conductors*.

When a metallic object is charged, the repulsive forces between the extra electrons drive them to the outer surfaces, so that the interior is left electrically neutral. In a demonstration of this effect, an instructor will charge a hollow conductor, such as a metal pail, by touching the inside with

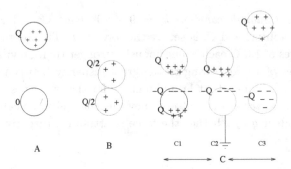

Fig. 12.4   A: A charged and an uncharged sphere; B: Charging by conduction, C: Charging by induction. C1: A charge is brought close, C2: The sphere is grounded, C3: The charging object is removed.

a charged rod. An electroscope is used to explore the inside: the charge has vanished! The electroscope then finds the missing charge on the outside of the pail. Similar effects are produced from an excess of positive charge, i.e. a deficit of electrons. So, what is an *electroscope*? It is an instrument to qualitatively measure the charge of an object. It consists of two thin metallic strips that are hinged at the upper end and can separate at the lower end. When it acquires a charge, the leaves separate and separation is a measure of the charge.

When a charged conductor is touched to an uncharged one, the charges are shared between the two conductors. If the conductors are identical in size, then each one gets half of the initial charge. See Fig. 12.4A and B. A conductor can also be charged by *induction*. In Fig. 12.4C a positive charge is brought close to an uncharged sphere (Fig. 12.4C1). The negative charges are attracted to the charging object and are held at the surface by the Coulomb attraction. The positive charges migrate to the far end. The sphere is now grounded and the positive charges are neutralized by the electrons from the ground (Fig. 12.4C2). The grounding wire carries these electrons. If the charging object is now removed, the negative charges spread out over the surface of the sphere, which has acquired a charge $-Q$ (Fig. 12.4C3)

12.4.1   *How does rubbing cause electrification?*

Static charging is caused by the transfer of electrons between two objects. The transfer is caused by the electrons' preference for one material over another, a kind of weak chemical bond, called *electronegativity*. Rubbing is not always necessary; close contact may be all that is needed (as you may have noticed when you unpack a box containing Styrofoam "peanuts"). But most objects have a coating of grime or oxide, so rubbing wipes away the grime and makes good electrical contact.

12.5 Electric Fields

How does the Coulomb force reach from one object to another? How can one understand *action at a distance*? We can ask the same question about gravitation, but we are so accustomed to living within the pull of gravity that we don't question how the force reaches us. To answer the question about electrical forces, Michael Faraday proposed that each charge carries an *electric field* **E** which radiates from the charge like light (Fig. 12.5). **E** is a vector; at every point it is equal to the force that would be felt by a unit positive charge, In the case of an isolated charge q, the magnitude of the field falls off with distance, as the inverse square

$$E = \frac{kq}{r^2}.\qquad(12.2)$$

For a positive charge the field vector points outward, away from the charge. For a negative charge the field points inward. *The field due to several charges is the vector sum of their individual fields.*

The force on a charge q' in a field **E** is

$$\mathbf{F} = q'\mathbf{E}.\qquad(12.3)$$

Although the electric field was a concept, invented to explain the force's action at a distance, we will learn in a later chapter, that the field has a very definite reality.

Illustrative Example 12.2: In the case of Illustrative Example 12.1, what is the electric field halfway between the two charges?

Fig. 12.5   The electric field lines around isolated electric charges.

Solution: If charge A is on the left of charge B, then the electric field of A is to the right and that due to B is to the left. If we call the first one positive, the second one is negative, and the total field is $kq_1/d^2 - kq_2/d^2$, where d is the distance of one of the charges to the center, namely 1 m. Thus the electric field is $\frac{9\times10^9\times(0.2-1.2)\text{C}\times10^{-6}}{1 \text{ m}^2} = -9 \times 10^3$ C/m^2.

12.6 Induced Polarization

Why do charges attract uncharged matter, such as a charged rod attracting a bit of paper?

   The electric field from the rod penetrates the paper; it extends through the atoms and affects their electrons and nuclei. Suppose the rod is charged positively, so that it attracts the electrons and repels the nuclei. In typical cases the forces are much weaker than those that hold the atom together, so they don't ionize them. But they distort the atoms; the electrons move slightly toward the rod, and the nuclei move away. The slight displacements of all the electrons and nuclei add up to a very small shift of a very large amount of negative charge toward the rod, and a small shift of positive charge away: the matter has become *polarized*. And because the shift is always such that the attracted charges have moved closer to the rod, and the repelled charges further away, the induced polarization always results in an attraction, whether the external charges are positive or negative (Fig. 12.6).

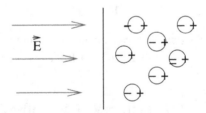

Fig. 12.6 An electric field induces dipole-like distortions in the molecular structure of an insulating material.

12.7 Electric Potential

When a charge moves in an electric field, it entails *work*, and a change in potential energy ΔU. For a constant electric field $\vec{E}$ and therefore constant force

$$\Delta U = Fd\cos\theta = qEd\cos\theta \,, \tag{12.4}$$

where θ is the angle between **E** and the displacement (distance moved) **d**. If $\vec{E}$ is in the direction of the displacement $\vec{d}$, then $\Delta U > 0$; if it is opposed, then $\Delta U < 0$. If it is perpendicular to $\vec{d}$, the $\Delta U = 0$.

There is a special term, $\Delta V = $ *electric potential difference*, the change in potential energy divided by the charge:

$$\Delta V = \Delta U/q = Ed\cos\theta \,. \tag{12.5}$$

The units of ΔV are Joules/Coulomb, equal to *volts*. Combining Eqs. (12.4) and (12.5),

$$\Delta U = q\Delta V \,. \tag{12.6}$$

12.8 Special Cases

The electrostatic potential or potential energy can be given in absolute terms for a point (very small) charge or a set of point charges because the potential an infinite distance away is zero. As for the case of gravitation, the potential is

$$V = kq/r \,, \tag{12.7}$$

and the potential energy for 2 charges q_1 and q_2 separated by a distance r is

$$U = kq_1q_2/r \,. \tag{12.8}$$

12.8.1 *Field between parallel plates*

A pair of parallel plates is separated by a distance d, and extend laterally very far compared to d. They have equal and opposite charges, which spread uniformly across the plates. (They are so wide that we need not consider effects at their edges). Because the charges are spread uniformly, the field has no component parallel to the plane of the plates; it must be normal to the plane and uniform everywhere across the plates. Let the upper plate be positive, **E** points down. The force on a unit positive charge is therefore **E** everywhere in the space between the plates. The work that must be done to bring a unit positive charge from the bottom to the top

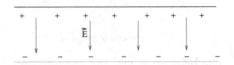

Fig. 12.7   The electric field between two large parallel conducting plates is uniform and perpendicular to the surface, when the spacing is small compared to the lateral extent of the plates.

plate is $\Delta V = Ed$; it is equal to the difference of electric potential, the *voltage* of the top plate relative to the bottom (Fig. 12.7).

Illustrative Example 12.3: Two parallel plates of large area are a distance $d = 1.2$ cm apart and the top plate is at a potential of 2.4 V relative to the bottom one. (a) What is the electric field between the plates. (b) If a 2 μC charge is placed a distance of 0.2 cm above the bottom plate, what is the force on it?

Solution: The electric field $|\vec{E}|$ is $\Delta V/d = 2.4$ V/1.2 cm $= 2$ V/cm. Since the field is constant and uniform, the electric field is the same throughout and it makes no difference where the charge is placed. The force on the charge is $q\vec{E} = 2 \times 10^{-6}$ C $\times 2$ N/C $= 4 \times 10^{-6}$ N.

Dipole field

Two equal and opposite charges a short distance apart form an electric *dipole*. Figure 12.8 illustrates a dipole's field lines. Some molecules have dipolar fields due to the arrangements of their constituent charges.

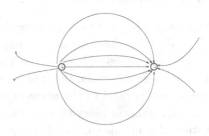

Fig. 12.8   Electric field lines around an electric dipole are symmetrical about a line going from one charge to the other.

12.8.2 *Field from a pointed conductor*

The field lines emerging from the surface of a conductor are perpendicular to the surface. The reason is that any field components that were initially parallel to the surface would cause charges to move, until those components would be reduced to zero. Thus, the field at the surface of a bump or pimple will radiate perpendicular to the bump. The field is more intense where the field lines are closer together, hence the field at the surface of a spherical tip of radius R varies in strength as $1/R^2$.

12.8.3 *Capacitance*

The ability of an object to store charge is called its *capacitance*, denoted by C, and measured in *farads*. When a voltage difference V is applied to a *capacitor*, the stored charge is

$$q = CV \,, \tag{12.9}$$

where q is in Coulombs, C in Farads, and V in Volts. An example of a simple capacitor is a pair of parallel plates. Its capacitance is proportional to the plate area and inversely proportional to the spacing; $C \propto A/d$. The typical capacitance values of capacitors in electronic circuits and apparatus are much smaller than one farad; they may be on the order of microfarads (μF) or smaller.

12.9 Field Ionization

The bonds that hold electrons in their atoms have limited strength, and can be ruptured by a sufficiently strong electric field. The field near a charged pointed conductor may be strong enough to ionize nearby atoms in the surrounding gas. The freed electrons and the ions are accelerated in opposite directions, and their speeds may be enough to ionize other atoms. *Field ionization* can be seen as glowing and dancing tendrils branching off the points of charged bodies. It is the "St. Elmo's Fire" that plays about ships' rigging and airplane wings when they are in a storm, and it is the electrical displays in "mad scientist" movies. The intensification of electric field at a point is the principal mechanism in Franklin's demonstration that lightning is an electrical discharge.

12.10   Some Applications of Electrostatics

There are many applications of electrostatics that are used in everyday life.

Variable capacitors, the plates of which mesh to a varying degree, can be used to tune a radio.

Coal plants make use of the attraction of light particles by charged objects to precipitate coal dust and remove them from the air.

Computer keyboards may use capacitors to recognize letters of the alphabet. The keyboard's keys are attached to a plunger that moves towards a fixed plate. As the plunger gets closer to the fixed plate the capacitance increases and the computer recognizes the change in capacitance to discern the letter that is being typed.

Copiers may use electrostatics. Selenium is an insulator in the dark, but turns to a conductor in light. The charge on selenium will remain there until it is illuminated. The drums on a Xerox copier may be made of Aluminum, but they are coated with Selenium.

Questions

Q12.1. When two objects of different materials are rubbed together there will be some charge transfer due to the difference in electron affinity. Then how can one account for the fact that sometimes your shirt or sweater will cling to itself as you take it off?

Q12.2. Why do the positive and negative charges of an object tend to be in balance?

Q12.3. Is it possible that the force that holds the planets in their orbits about the Sun is actually electrostatic rather than gravitational?

Q12.4. Why does the charge on a metal object reside on its outside surface?

Q12.5. Why is the charge on a flat metal plate distributed uniformly over its surface?

Q12.6. Why does grounding an object discharge it?

Q12.7. In charging by induction, you keep the charging object close by until the last step. What happens if you remove it earlier (say at Fig. 12.4C2)?

Q12.8. Bits of paper are attracted to a charged rod. Why?

Q12.9. In a region of space where the charge is zero, is $\vec{E}$ necessarily zero? Explain.

Q12.10. If you double the separation between two charges, by how much does the force change? If you double both charges, by how much does it change?

Q12.11. A negative charge is at rest. If an electric field is introduced, will the charge move along the electric field or opposite to it? Why?

Q12.12. If a positive charge q is moved towards a negative charge, will the potential energy of the positive charge increase, decrease, or remain the same? Why?

Q12.13. If a negative charge, $-q$, is moved in the direction of an electric field, will its potential energy increase, decrease, or remain the same? Why?

Q12.14. In a lightning storm, is it better to stand on a conductor or insulator? Why?

Q12.15. Why is the electric field on the surface of a charged conductor perpendicaular to the surface?

Q12.16. If you move a positive charge towards a negative one, does the potential energy of the positive charge increase, decrease, or remain the same? Why?

Problems   *Please provide 2 significant figures for answers*

P12.1. A glass rod, after rubbing with a silk cloth, has lost 1.0×10^{10} electrons. What is the charge on the rod, in Coulombs? What is the charge on the silk? (The charge of an electron is -1.60×10^{-19} C).

P12.2. Estimate the Coulomb force between you and your classmate across the aisle, if you transferred 1 percent of your electrons to your neighbor. Assume that you are primarily composed of water. 1 gram of water contains $\approx 1.0 \times 10^{23}$ electrons, and each electron has a charge -1.60×10^{-19} C. Compare this force to your gravitational attraction.

P12.3. A hollow metal sphere has a small hole that permits a charged rod to be inserted into the cavity. The outer diameter of the sphere is 1.0 meter, and its inner diameter is 0.5 meter. A charge of $+1.0 \times 10^{-5}$ C is placed on its inner surface. How much force is exerted on a similarly charged sphere of the same dimensions, 10.0 meters away?

P12.4. Two positive charges are located on the x-axis. The smaller one of 0.20 C is at $x = 1.5$ m; the larger one of 0.30 C is located at $x = -2.5$ m.
(a) Find the electric field at the origin.
(b) Is there an "equilibrium" position where $\vec{E} = 0$? If so, locate it.
(c) Is the equilibrium position stable? That is, if you move to the left or to the right, or up or down from it, does the force tend to return a charge to the equilibrium position?

P12.5. Charges $+1.0$ μC and $+3.0$ μC are 1 meter apart along the x axis.
(a) Where should a test charge $+0.5$ μC be placed so that it would experience zero net force?
(b) Is it stable in this position?
(c) Repeat parts a and b, assuming that the test charge is negative.

P12.6. For Problem 12.4,
(a) how much work is require to move the smaller charge to the origin? Assume a constant electric field equal to that at 0.75.
(b) What is the change in potential energy and potential for this move?
(c) What would be the answer to (b) if the move were by the same distance but in the opposite direction ($x = 1.5$ m to $x = 3$ m)? Assume the same electric field. (Note: Be careful with signs).

P12.7. Four equal charges, q, are located at the corners of a square at $x = y = \pm 2.5$ cm.
(a) Where is the electric field $= 0$?
(b) Is this equilibrium position stable? That is, does the potential energy increase in all directions?
(c) What is the potential at the origin?
(d) How much potential would be gained or lost at the origin if one of the charges is removed to infinity, where the potential is zero?

P12.8. A pair of parallel plates are 0.5×10^{-4} m apart. They are connected to a 90 V battery, with the + terminal connected to the top plate.
(a) How large is the field between the plates?

(b) What is the force on an electron between the plates?

(c) When the electron moves from the bottom to the top plate, how much energy does it gain or lose?

P12.9. How much charge is stored in a 0.05 μF capacitor connected to a 10 V battery?

P12.10. A capacitor is connected to a 6.0 V battery and has a charge of 15 μC. What is its capacitance?

P12.11. (Challenge) An electron traveling at 3.0×10^6 m/s in the x direction, enters a region where there is an electric field of 1.0×10^3 volts/m pointing upward in the y direction. The electron enters at $y = 0$. After the electron has passed through a 0.25 meter extent of the field, in what direction is the electron moving?

P12.12. (a) Two positive charges of charge q and $2q$ are placed at the origin and at 2.0 cm, respectively. Where is the electric field zero?

(b) If the charge q at the origin is replaced by a charge $-q$, where is $\vec{E} = 0$?

Answers to odd-numbered questions

Q1. The charging is patchy. An uncharged part of the sweater is inductively charged when it comes in contact with a charged part. Then the two parts are attracted to each other. The effect occurs independent of polarity.

Q3. No. There would be repulsive forces between the planets, contrary to evidence.

Q5. The charges repel each other, and spread out uniformly on the surface; the charge density (charge per unit area) is lowered as far as it can be.

Q7. No charge will be transferred. The charges go to ground.

Q9. No, because there may be an electric field due to charges elsewhere.

Q11. It will move anti-parallel to $\vec{E}$, since $\vec{F} = q\vec{E}$.

Q13. If a negative charge is moved parallel to $\vec{E}$, its potential energy increases, since work is required.

Q15. Any electric field parallel to the surface would redistribute the charge until the parallel component of the field vanishes.

Answers to odd-numbered problems

P1. $Q = 1.6 \times 10^{-9}$ C. Some electrons were transferred from the glass to the silk; the charge on the silk is negative.

P3. The charge spreads out uniformly on the outer surface. Like all spherical charge distributions, the charge produces a field that acts as if the charge is concentrated at the center. The force is 9×10^{-3} N.

P5. (a) The ratio of the distances between the test charge and the two fixed charges should be $+1/(1 + \sqrt{3})\mu$C.
(b) It would not be stable against displacements in the y direction.
(c) The ratio of distances would be the same. The position of the test charge would be unstable in the x direction.

P7. (a) $\vec{E} = 0$ at the origin.
(b) It is an unstable equilibrium.

(c) $V = 4$ kq/3.5 m.
(d) $\Delta V = $ kq/3.5 m.

P9. $q = CV = 0.5$ μC.

P11. The force on the electron is $F_y = 1.6 \times 10^{-16}$ N, causing an acceleration $a_y = -qE_y/m = -1.75$ (≈ 1.8) $\times 10^{14}$ m/s. During its transit time $t = 0.25$ m/(3×10^7) m/s $= 8.3 \times 10^{-9}$ s it acquires an upward velocity $v_y = -1.46 \times 10^6$ m/s. The velocity vector is pointing at an angle $\tan^{-1}(v_y/v_x) = \tan^{-1} 1.46/3 = 26°$ downward from horizontal.

Chapter 13

Currents and Circuits

13.1 Power Sources

13.1.1 *Household power*

Static electricity is fine for demonstrations, but you can't use it to run a laptop, or any other appliance. What you need is a source of continuous and robust electrical power. That's what you get when you plug into an electrical wall socket; 110 Volts AC. "110 volts" is the magnitude of the *electromotive force*, abbreviated *emf*, symbol $\mathcal{E}$. "AC" stands for "alternating current"; the *voltage* oscillates in time, as in Fig. 13.1, alternating from positive to negative at a frequency $f = 60$ Hz (sometimes called "60 cycles". In some European countries the emf frequency is 50 Hz).

$$\mathcal{E}(t) = \mathcal{E}_p \sin(2\pi f t), \qquad (13.1)$$

The peak amplitude $\mathcal{E}_p = 110\sqrt{2}$ volts; we'll explain later where the $\sqrt{2}$ comes from.

13.1.2 *Batteries*

Another source of emf is a battery. It delivers DC ... "direct current", a steady emf, from a source of chemical energy. Batteries were invented by Alessandro Volta in the mid 1700's in Italy. His invention followed the work of Luigi Galvani, a physician, who showed that he could produce electrical effects from frogs' legs.

The two electrodes of a battery are composed of two different materials, usually two metals or metallic compounds, which have different *electro-chemical potentials*, which measure their tendency to dissolve atoms into

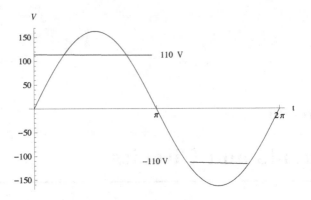

Fig. 13.1   60 Hz alternating voltage.

an *electrolytic* solution. As it dissolves, each atom leaves one or two electrons on the *anode* — the negative terminal — and enters the solution as a positive ion. The electrolyte is a liquid or paste. It is electrically conducting, and fills the space between the electrodes (Fig. 13.2). The anode is surrounded by an "atmosphere" of positive ions in the electrolyte. When a current is drawn from the battery (it is the electrons which "flow" through the circuit) the external circuit takes negative charge away from the negative terminal. This allows some of the positive ions to drift away through the electrolyte to the *cathode* — the positive terminal. As the "atmosphere" is depleted, the anode "pumps" more ions into the electrolyte. The "pump" is driven by chemical potential energy, and this energy is expended as current is drawn, by losing material from the electrode. As the battery nears the end of its useful life the accumulation of a non-conducting coating (an oxide or other compound) on the electrodes increases the battery's internal resistance.

The typical emf produced by a single "cell" ranges from a little less than 1 V to about 2 V, depending on the compositions of the electrodes and the electrolytic medium. An example of a typical "dry cell" is shown in Fig. 13.3. Type AAA, AA and C batteries, have a metal button on one end, the positive *electrode*. The metal bottom of the case is the negative electrode. Battery emf's range from about 1 volt upward. The physical size of the battery is a rough guide of its useful life, meaning how much current can be drawn for how much time: its capacity in "ampere-hours". One AAA battery can supply 1 ampere current for about one hour; its capacity is ≈1 A-hr. Some batteries have voltages as high as 45 V; they are composed of

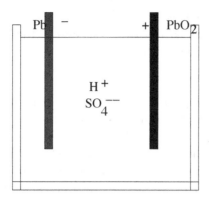

Fig. 13.2 Picture of a lead-acid battery. At the negative terminal $Pb + SO_4^{--} \rightarrow PbSO_4 + 2e^-$; at the positive terminal $PbO_2 + SO_4^{--} + 4H^+ + 2e^- \rightarrow PbSO_4 + 2H_2O$.

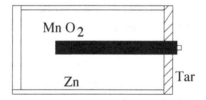

Fig. 13.3   Picture of a typical dry cell.

several cells, internally connected in one package. You can connect several batteries in series, negative terminal to positive terminal of the previous one, or in opposition. In series, the emf's add; in opposition, they subtract. If you connect several batteries of the same voltage in parallel, you boost the power output.

13.2   Electrical Current and Resistance

Electrical current is equal to the rate of passage of charge through a wire or a region of space:

$$I = \Delta q / \Delta t. \tag{13.2}$$

The unit of current is the *Ampere*, abbreviated A; 1 Ampere = 1 Coulomb/second.

In ordinary conductors (not *superconductors*, which we get to else-where), the wire has an *electrical resistance* R that impedes current flow. *Ohm's Law* relates current to the voltage:

$$I = V/R. \tag{13.3}$$

The law was propounded by Georg Ohm, a 19th century high school teacher and university lecturer. Ohm gave this very simple relationship on the basis of careful and extensive measurements. At the time it was not known how electrical effects were transmitted. Ohm's view that the current was a flow of a material substance seemed to be an original idea, although Benjamin Franklin had deduced that from his experiments on electrostatics.

The unit of resistance R is the *ohm*, and the standard symbol is the Greek upper case Omega, Ω; dimensionally, 1 ohm = 1 volt/1 ampere. If an emf of 100 V is connected to a wire that has a resistance of 10 Ω, it will cause a current of 10 A.

Electrical resistance of a wire depends on its material; it is proportional to its length L, and inversely proportional to its cross sectional area a:

$$R = \rho \frac{L}{a}, \tag{13.4}$$

where ρ is the *resistivity* of the material. The unit of resistivity is *ohm-m* or *ohm-cm*. A wire of low resistivity is generally a pure metal such as copper and aluminum. The atomic structure is *crystalline*: the positive ions are arranged in orderly rows, forming simple channels for the electrons to flow through, with little interference. For a good conductor like copper, the resistivity is small, of the order of 1.7×10^{-8} ohm-m. In contrast, a metal with high resistivity is disordered; it is an *alloy*, composed of a two or more elements, which form irregular arrangements. An alloy of nickel, chromium and other metals, tradenamed "Nichrome", is a high resistivity metal used in toasters, hair driers, and other heating appliances. Its resistivity is about 100 Ω − cm (note the cm, not m).

13.2.1 *A microscopic description of an electric current*

We imagine a super-magnified view of an electric wire, so that we can travel along with the electrons. The space around us is filled with ions that look like fuzzy balls; they have an ordered arrangement, like the oranges on display in a market, but they are in constant motion about their average positions, vibrating and bouncing against each other. Moving between them are the conduction electrons, which keep to the interstices between

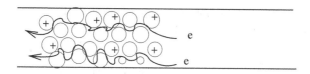

Fig. 13.4 Schematic view inside a wire. Electrons move through a lattice of positive ions.

the ions. The electrons are jittering very rapidly, but also drifting in one general direction. As they move, they occasionally collide violently against an ion. The collision knocks the electron off course and takes some of its kinetic energy, which makes the ion vibrate more vigorously (Fig. 13.4). Thus, the collisions cause the electrical resistance, and the electrons' energy losses are turned into heat. The drift velocity of the electrons along the wire is much smaller than their actual velocity.

The flow of electrical charge is akin to the flow of a liquid (e.g., water), and the analogy can be useful. The rate of water flow is measured in volume/second, that of charge in Coulombs/second, which is equivalent to current. The pump or whatever starts water flow is the battery in an electrical circuit; the pipes are the wires, and if there is a valve to regulate the flow, it is replaced by a switch in an electric circuit.

13.3 Electrical Power

When a source of emf drives a current through a circuit it is expending energy at a rate equal to the product of voltage and current:

$$P = \Delta E / \Delta t = V I \,. \tag{13.5}$$

The unit of electrical power is the same as for mechanical power, the *Watt*, abbreviated W. The product of x Volts and y Amperes is xy Watts. 1 W = 1 V $\times$ 1 A.

We can use Ohm's Law to express the power in alternative forms to Eq. (13.4):

$$P = VI = V^2/R = I^2 R \,. \tag{13.6}$$

$I^2 R$ is the *resistive loss; it is the rate of conversion from electrical energy into heat.*

$$\frac{[V(t)]^2}{R} = \frac{V_p^2}{R}[\sin(2\pi f t)]^2 \,.$$

What is more important, for cost and heating effect, is the *average power*. The average value of $\sin^2(2\pi ft)$ (and of $\cos^2(2\pi ft)$ over a full cycle is $\frac{1}{2}$, therefore

$$(VI)_{av} = \frac{V_p^2}{2R} = \frac{V_{eff}^2}{R}. \tag{13.7}$$

Equation (13.7) shows that the *effective value* of the voltage is

$$V_{eff} = V_p/\sqrt{2}.$$

This is the origin of the factor $\sqrt{2}$, which is mentioned at the beginning of the chapter.

Illustrative Example 13.1: A copper wire of length 2.0 m and diameter 2.0 cm is used to connect a household circuit (120 V AC) to a 60 W lamp.
(a) Find the resistance of the connecting wire.
(b) Find the resistance of the lamp.
(c) Neglecting any loss of power in the wire, what is the current delivered to the lamp?

Answer: (a) R of wire $= \rho L/A$. This is $1.7 \times 10^{-8}\,\Omega - m \times 2\,m/\pi(10^{-4})m^2 = 1.1 \times 10^{-4}\,\Omega$.
(b) R of lamp $= V^2/P = (120)^2/60 = 240\,\Omega$.
(c) $I = V/R = 120V/240\,\Omega = 1/2$ A. To check, $P = I^2R$ or $I^2 = P/R = 60\,W/240\Omega = 1/4\,A^2$ and $I = 1/2$ A.

13.4   Lamps, Circuit Breakers, Switches

Of all the possible uses for electricity, the simplest is to generate heat, but it is a very inefficient use of energy. Most of the electric power produced in the United States begins with burning a fuel to run an electric generator, then the power is delivered over long distances via high voltage transmission lines, and finally converted to lower voltage by transformers. (This will be explained in a later chapter). There are inefficiencies at each stage of the process. However, electrical energy is very convenient, and might be the only practical way to deliver heat where it is needed, such as the various domestic heating appliances: clothes dryer, hot water tank, oven and cook-top, toaster and hair dryer. Many homes are heated by electricity; it would be more energy efficient to replace it with oil or gas heating if they are available.

A substantial fraction of the electric power consumed in the home is still used for lighting. A light bulb, which was common some years ago, or an *incandescent lamp* produces light by heating a wire filament to such high temperature that it is "white hot". In a typical bulb, most of the energy is radiated as heat; as little as fifteen to twenty percent of the energy is emitted as visible light.

Fluorescent lamps are much more efficient light producers. Their basic operating principle is a *gas discharge*. Typically, it is a sealed glass tube containing a low pressure gas such as argon, with a small concentration of mercury vapor. The discharge is started by a temporary heating of a filament or a high voltage arc to supply a stream of free electrons, then the discharge is maintained by ionizing some of the gas molecules. The radiation from excited atoms is mainly in the ultraviolet, well beyond the visible part of the spectrum. Therefore, the tube is internally coated with a *phosphor*, which converts the radiation to visible light. More recently, there has also been the introduction of LEDs or *light-emitting diodes*; they use electrons and "holes" (missing electrons) in a semi-conductor. Ohm's law does not apply to them. Examples of semi-conductors are C, Ge, and Si. Their resistance varies with voltage. For small voltages, they are insulators; above a critical voltage they become conductors because some atoms are ionized (electrons are removed), giving rise to free electrons. Solar cells, which convert light to electricity, also make use of semi-conductors.

A *fuse* is inserted into a circuit to protect it against excessive current. The fuse is made of a short and thin metal wire and has a resistance such that a large current heats the wire to its melting point, thereby breaking the connection. Fuses have been used in household wiring and in some electrical appliances, but they are now obsolescent, and have been largely replaced by *circuit breakers*, which are magnetic devices that can be reset after the cause of the excessive current has been removed.

Switches are used to turn devices (e.g. lamps) on and off. They are simple and break the circuit, which prompts the current to drop to zero, or turn on the device by closing the circuit.

13.5 Series and Parallel Connections

The various appliances in the home are all designed to operate independently, each being connected to the household AC power. They are connected *in parallel*, and this is schematically shown in Fig. 13.5(b).

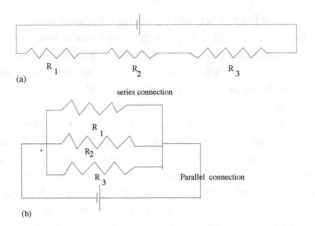

Fig. 13.5   Series and parallel connections of resistances.

In a *series connection*, the same current flows through each resistor in turn, and the resistance R_s of the combination is the sum R_s (s for series)

$$R_s = R_1 + R_2 + R_3 + \cdots \qquad (13.8)$$

The voltage across any one of the 3 resistances in series shown in Fig. 13.5(a), say R_1, is $V_1 = IR_1$, where I is the current.

In parallel circuits the same emf $\mathcal{E}$ is connected to all of the appliances, shown as resistances in Fig. 13.5(b). The current divides among the parallel branches, with the smallest resistance taking the most current. These branch currents add up to the total that flows through the source. The current through any one of the resistors, say R_1 is

$$I_1 = \mathcal{E}/R_1 , \qquad (13.9)$$

if the resistances are connected to an emf $\mathcal{E}$. The resistance of a combination of resistors connected in parallel is lower than the resistance of any one of them. The resistance R_p (p for parallel) of the combination is determined by adding the currents: $I_{\text{total}} = (I_1 + I_2 + I_3 + \cdots)$; then since $I_1 = \mathcal{E}/R_\infty$ etc., we get:

$$\frac{1}{R_p} = \frac{1}{R_1} + \frac{1}{R_2} + \frac{1}{R_3} + \cdots \qquad (13.10)$$

The analogy to water is useful. A series circuit is like water going through a variety of connected pipes of different diameter. The total flow of water does not change, but the velocity may. A parallel circuit is like one pipe dividing into several ones, so that the total water volume divides

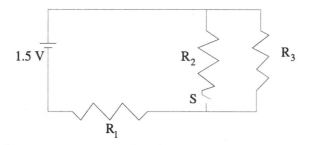

Fig. 13.6   Circuit of series and parallel resistances.

between the various pipes. If these pipes are joined again, the flow of water becomes identical to that before the division. Similarly, the current in Fig. 13.4(b) divides among the three resistances and combines to the same current when the resistances are joined again.

How should you tackle the analysis when the circuit is a combination of series and parallel pieces? Treat the internal combinations (pieces) first. Calculate the resistance R_c of each combination (whether series or parallel), then redraw the circuit with the R_c's for the combinations. You should end up with a circuit with only series resistances. Add these resistances to find the total resistance. You can then find the total current. The current is the same for all the series resistances you have found. In Fig. 13.6, for example, you first find the resistance, R_{23} for the combination of R_2 and R_3; then $R_{tot} = R_1 + R_{23}$. Once you have the total resistance, you can find the total current.

Illustrative Example 13.2: Consider $R_2 = 3\ \Omega$, and $R_3 = 6\ \Omega$ connected in parallel and a resistance $R_1 = 2\Omega$ connected in series with these two, as in Fig. 13.6. The entirety is connected to a 1.6 V battery.
(a) Find the total resistance of the circuit.
(b) Find the current through the 2 Ω resistance and the 3 Ω resistance.
(c) Find the voltage across the 2 Ω resistance.
(d) Find the power used by the circuit.

Answer: The resistance of the two parallel resistors is $R_2 R_3/(R_2 + R_3) = 18/9 = 2\ \Omega$.
(a) The total resistance is $R_t = 2 + 2 = 4\ \Omega$.
(b) $I_1 = I_t = 1.6\ \text{V}/4\ \Omega = 0.4\ \text{A}$. The current throughout the 3 Ω resistance

can be found by first finding the voltage across it. This is $1.6 \text{ V} - 2 \, \Omega \times 0.4 \text{ A} = 0.8 \text{ V}$. Thus, $I_2 = 0.8 \text{ V}/3 \, \Omega = 0.27 \text{ A} \approx 0.3 \text{ A}$.

(c) $V_1 = I_1 R_1 = 0.4 \times 2 \, \Omega = 0.8 \text{ V}$.

(d) $P = IV = 0.4 \text{ A} \times 1.6 \text{ V} = 0.64 \text{ W} \approx 0.6 \text{ W}$.

13.6 Semiconductors, Transistors and Computer Chips

A semiconductor's resistance varies with the applied voltage. For small voltages it is an insulator; for voltages above a threshold value V_o, the resistance decreases. The cause is that some atoms become ionized by the higher voltage, a process of internal "field ionization" that liberates electrons for conduction (see Sec. 12.10). A semiconductor can be fashioned into a "transistor", to control the current in a circuit, e.g. to *amplify* a weak current into a larger one. In 1947, Bell Labs scientists John Bardeen, Walter Brattain, and William Shockley created a device that evolved into the modern transistor, for which they earned the 1956 Nobel Prize for Physics. It has been called "the most important invention of the 20th century".

Transistors and other circuit elements such as resistances and capacitors can be "printed" on the surface of "substrates" of silicon or germanium by *molecular beam evaporation*. Modern fabrication methods can produce computer chips whose surfaces have circuits composed of thousands of tiny transistors and other circuit elements. Chips for specialized uses are the "brains" of computers, calculators, communications equipment, appliances and toys.

Questions

Q13.1. In a series circuit, the current flows through each of the resistors in turn. Does the current get weaker as it flows though each resistor? Explain.

Q13.2. An AC emf is connected to a resistor. As the voltage alternates from $+$ to $-$, how does the heat dissipation vary with time? Sketch the function.

Q13.3. A battery with an emf of 6 V is in series with three 100 ohm resistors connected in parallel. How much current flows through one of the resistors?

Q13.4. What are the disadvantages of a series versus a parallel resistance circuit?

Q13.5. In a circuit of a battery and a lamp, is the current leaving the light bulb the same as that entering it? Explain.

Q13.6. In Fig. 13.5(b), if the parallel resistances are in a circuit, what happens to the total current when R_3 is disconnected? Explain.

Q13.7. How would you connect an ammeter to read the current in R_1 in Fig. 13.5(b)? Explain.

Q13.8. How would you connect a voltmeter to read the voltage lost in R_1 in a circuit with the series resistances of Fig. 13.5(a)? Explain.

Q13.9. In the series resistances shown in Fig. 13.5(a), assume that $R_1 > R_2 > R_3$. Which resistance will have the largest current when the resistances are connected to a battery with the + terminal closest to R_1? Which one will have the largest voltage drop? Explain.

Q13.10. Repeat Q13.9 for the parallel resistances of Fig. 13.5(b).

Problems   *Please give answers to 2 significant figures.*

P13.1. How could four 100 ohm resistors be connected to give a circuit resistance of 250 ohms?

P13.2. What is the resistance of a 100 Watt light bulb? (100 W is the power level when connected to a standard household power outlet).

P13.3. A calculator is designed to take three 1.2 V batteries connected in series. What is the operating voltage of the calculator? Sketch how the batteries should be connected.

P13.4. A series-parallel circuit is sketched in Fig. 13.6. The resistances are $R_1 = 3\ \Omega$, $R_2 = 4\ \Omega$, $R_3 = 7\ \Omega$.
(a) Calculate the branch currents I_1, I_2, I_3 when the switch S is open.
(b) Calculate the branch currents when the switch is closed.

P13.5. For the series resistances of Fig. 13.5(a), the three resistances are $6\ \Omega$, $15\ \Omega$ and $12\ \Omega$. Find the total resistance and the current if the circuit is connected to a 9 V battery.

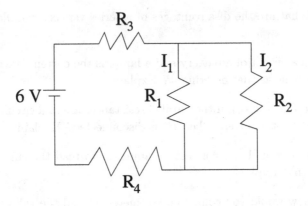

Fig. 13.7   Circuit of series and parallel resistances.

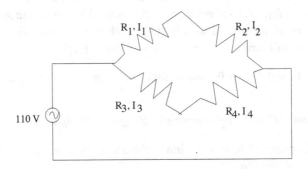

Fig. 13.8   Circuit of series and parallel resistances.

P13.6. Repeat P13.5 for the parallel resistances of Fig. 13.5(b). Find the current in the 15 ohm resistance.

P13.7. For the circuit shown in Fig. 13.7, find the currents I_1 and I_2. The resistances are: $R_1 = 6\ \Omega$, $R_2 = 10\ \Omega$, $R_3 = 2\ \Omega$, $R_4 = 5\ \Omega$.

P13.8. For the circuit shown in Fig. 13.8, find I_1, I_3, V_1, V_2, V_3, V_4. The emf is 110 V AC and the resistances are $R_1 = 6\ \Omega$, $R_2 = 9\ \Omega$, $R_3 = 2\ \Omega$, $R_4 = 6\ \Omega$.

P13.9. For the circuit of P13.4, find the voltage across each resistance when the switch S is closed.

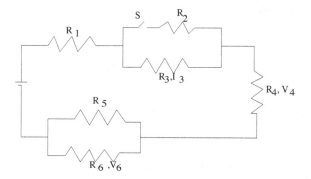

Fig. 13.9   Circuit of series and parallel resistances.

P13.10. For the circuit shown in Fig. 13.9, the emf is 6 V and the resistances are $R_1 = 1\ \Omega$, $R_2 = 3\ \Omega$, $R_3 = 5\ \Omega$, $R_4 = 2\ \Omega$, $R_5 = 6\ \Omega$, $R_6 = 8\ \Omega$. Find the following quantities when the switch is open and when it is closed:

(a) R_t, the total resistance of the circuit.

(b) V_4, V_6.

(c) I_1, I_3, I_4, I_5.

(d) The total current.

Answers to odd-numbered questions

Q1. The current does not get weaker, since the number of electrons moving out of a region must equal the number moving in.

Q3. Since the resistors are in series, the total circuit resistance is 300 ohms, thus the current through each resistor is the same, $6/(300) = 0.02$ A.

Q5. The current is the same, since none is lost.

Q7. You would connect the ammeter in series with the resistance, since it measures the flow of charge.

Q9. I is the same. V_1 is the largest $= IR_1$.

Answers to odd-numbered problems

P1. Two in series, plus two in parallel.

P3. 3.6 V, connected from the case of one battery to the positive terminal of the next.

P5. The resistance is $(6 + 15 + 12)$ ohms $= 33$ ohms. Therefore the current is 6 V/33 ohms $= 0.27$ A.

P7. $R_t = 30/8 + 7 = 10.8 \ \Omega$. $I_t = 6 \text{ V}/10.8 \ \Omega = 0.56$ A; $I_1 = (R_2/R_p)I_t = (5/8)0.56 = 0.35$ A; $I_2 = (R_1/R_p)I_t = (3/8)0.56 = 0.21$ A.

P9. $V_1 = 6.2$ V, $V_2 = V_3 = 8.3$ V.

Chapter 14

Magnetism

14.1 Beginnings of the Study of Magnetism

In Greek mythology, a mountain in the region of Magnesia had the power of drawing the iron nails out of a walker's shoes. Only a myth, but its probable basis was an outcropping of *lodestone* ("stone that leads" referring to the magnetic compass), a naturally magnetic rock. It is an iron ore, ferric oxide Fe_2O_3, called *magnetite*.

Magnets and magnetism have inspired wonder and "explanations" of their powers. There were fanciful theories by several philosophers, including Diogenes, who claimed, in 5th century BC Greece, that there is "humidity in iron, which the dryness of a magnet feeds upon". That superstition persisted into the sixteenth century CE. Serious study began with the work of William Gilbert, whose book *De Magnete* in 1600 was one of the principal events in the development of the scientific method. He ridiculed the meta-physicians, for not being practiced in the subjects of Nature, and being misled by certain false physical systems, they adopted as theirs, from books only, without magnetic experiments, certain inferences based on vain opinions, and many things that are not, dreaming old wives tales.

14.2 The Earth is a Magnet

Since Gilbert's pioneering work, magnetism has been studied intensively, for scientific interest and for its many practical applications. But long before Gilbert's research, pieces of magnetite had been used as a great aids to navigation. A *magnetic compass* is basically a small magnet that is free to pivot about a vertical axis. It tends to align itself approximately

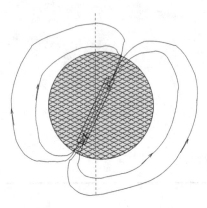

Fig. 14.1   The magnetic field of the earth is similar to an internal bar magnet, inclined relative to the rotation axis, with the magnetic S pole uppermost. The field is that of a magnetic dipole.

along the Earth's rotation axis, with one end of the magnet, the *North-seeking pole*, pointing in the general direction of the Earth's North Pole. This end of the magnet is called its *North pole*; the other end, its *South pole*. Magnets' ends are either S or N. And because *unlike poles attract*, the Earth's magnetism acts as if there is a great internal bar magnet, aligned approximately parallel to its axis, with its *S end closer to the North Pole*. See Fig. 14.1 for a sketch of the earth's magnetic field. Notice a difference between the electric and magnetic field lines. The electric field lines start on a positive charge and end on a negative one. By contrast, the magnetic field lines form complete loops; they start and end at the N pole. Outside the magnet they go from N to S poles, but inside the magnet they go from the S pole to the N pole.

Actually, there is no such internal bar magnet in the earth; the actual source of the Earth's magnetism is electric currents, discussed later in this chapter.

14.3   Basic Magnetism

Magnetism and static electricity are similar in several ways, but there are also some sharp contrasts. Consider a bar magnet, such as you may have had as a toy, or better yet, two bar magnets. They attract each other, S to N. S repels S and N repels N. Thus, like static electricity, unlike poles and charges attract; like poles repel. Also similar to electricity, the forces act along the straight lines between the poles, and vary in strength as the inverse square of the distance between them,

$$F \propto m_1 m_2 / r^2 \,, \tag{14.1}$$

where m_1 and m_2 are the pole strengths. The magnetic force is transmitted through a *magnetic field* **B**. B is the force on a unit N pole; the magnetic field unit is the *Tesla*, named in honor of the 20th century scientist and inventor, Nikola Tesla. (Read more about Tesla in Chapter 15).

In contrast to electricity, which can have isolated positive and negative charges, the magnetic N and S poles of a magnet cannot be separated from each other. If a magnet is broken in two, instead of separate N and S poles, we get two magnets, each with its own N and S ends. And if we break those halves, we end up with four magnets, all of them complete with N and S ends. There have been some very strenuous attempts to find or create magnetic *monopoles*, but all were unsuccessful: *There are no free magnetic poles; the basic magnetic objects are magnetic dipoles.*

The strength of a magnetic dipole is measured by its *magnetic moment* $\vec{\mu}$, which is the product of its pole strength and the separation distance between the poles, d: $\vec{\mu} = m\mathbf{d}$, where the direction of **d** is from S to N, (Fig. 14.2). The strength of a magnetic dipole, $|\vec{\mu}|$, can be measured in Joules/Tesla, where the magnetic field is measured in Teslas, abbreviated by T.

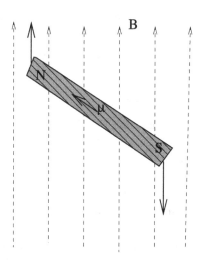

Fig. 14.2 A magnetic dipole in a magnetic field $\vec{B}$. The dipole tends to align along the direction of $\vec{B}$.

Note that the energy is negative if $\vec{\mu}$ is parallel to $\vec{B}$. The lowest (most negative) energy occurs when the dipole is aligned parallel to the magnetic field **B**; At an angle θ between $\vec{\mu}$ and **B** the magnetic energy is

$$E = -\mu B \cos\theta\,,$$
$$E = -\vec{\mu} \cdot \vec{B}\,. \tag{14.2}$$

The units of $\vec{\mu}$ are Joules/Tesla. The magnetic field exerts a torque on a magnetic dipole

$$\vec{\tau} = \vec{\mu}x\mathbf{B}\,, \tag{14.3}$$

where the symbol **x** denotes the *vector product* or *cross product*. The magnitude of the cross product **AxB** is $AB\sin\theta$; so here, it is $\mu B\sin\theta$; the direction of the torque is pointed by your thumb, when the fingers of your right hand curl *from* the direction of **A** *to* **B**.

Illustrative Example 14.1: Find the torque and energy for a bar magnet of pole strength $m = 0.41$ J/T-m and of length 3.0 cm, in a magnetic field $\vec{B}$ of 0.30 T at an angle of 30° relative to the magnetic dipole.

Solution: The magnetic dipole strength is $\mu = 0.41$ J/T-m $\times$ 0.03 m $= 1.2 \times 10^{-2}$ J/T.
The torque is $|\vec{\tau}| = \mu B\sin\theta = 1.2 \times 10^{-2}$ J/T $\times$ 0.30 T $\sin(30)° = 1.8 \times 10^{-3}$ J.
The energy is $E = \mu B\cos\theta = 1.2\times10^{-2}$ J/T$\times$0.3 T$\cos(30)° = 3.1\times10^{-3}$ J.

14.3.1 *Magnetic materials*

Magnets strongly attract materials that have a large *magnetic susceptibility*. Materials with high magnetic susceptibility (symbol χ) include iron, cobalt and some of their alloys.

The magnetic susceptibility of all materials is temperature dependent. At high temperatures, iron is not strongly affected by an external magnetic field. The attraction increases gradually as the iron is cooled; then at a definite temperature, its *ferromagnetic Curie temperature* T_c, the interaction begins a steep rise. The ferromagnetic Curie temperatures of iron and other materials are listed below. Dysprosium is one of a number of elements that become ferromagnetic when cooled far below room temperature.

Table 14.1 Magnetic materials.

Substance	Ferromagnetic Curie Temperature, T_c, °C
Iron	770
Cobalt	1127
Ferric oxide, Fe_2O_3	500
Nickel	358
Dysprosium	−168

14.3.2 *Magnetic domains and atomic magnetism*

An ordinary, unmagnetized piece of iron below T_c is composed of *magnetic domains*; magnetized regions of different orientations, so oriented that their magnetic fields cancel each other. If the iron is very pure, it has high *magnetic susceptibility*, the domains can be shifted (grow or diminish) in a relatively weak external field, and the object then acquires a net magnetic moment. When the external field is removed, the domains tend to relax back to their original states, so that their fields again cancel. In contrast, if the iron is *magnetically hard*; it requires a strong external field to affect the domains, and when the strong external field is removed, some effect is retained, so that the iron is converted into a *permanent magnet.*

The basic magnetism of the domains springs from the material's microscopic properties. The elements are its electrons, all of which have intrinsic magnetic moments in addition to their charge. One or a few of the electrons of each atom can be aligned in an external field, such as the field due to the neighboring atoms' electrons. At high temperature, thermal vibrations shake the electrons, so that they have random orientations. As temperature decreases the agitation weakens, and the electrons respond more readily to the magnetic fields produced by their neighbors. Small clusters of atoms begin to act as magnetic units, with a finite magnetic moment that fluctuates rapidly in direction. With continued cooling, the clusters grow in size and number, their fluctuations slow down, and each cluster, acting as a unit, senses the fields due to nearby clusters. Their interaction becomes stronger, until at some temperature, the *ferromagnetic Curie temperature* the clusters lock together, and form macroscopic domains.

14.3.3 *Magnetic recording*

Magnetic recording tapes and disks have coatings of a ferromagnetic powder such as iron oxide. A *recording head*, located very close to the oxide surface, locally magnetizes the material by currents from an amplifier, as the tape is moving past. The amplifier output, which is either digital or analog, is driven by the input voice, music, or data source. The oxide coating is magnetically hard, so that the magnetization becomes a permanent record. When the recorded tape is played back, the changing magnetic field induces an electric signal which is then amplified by an electric circuit.

14.4   Magnetic circuits

A magnetic field is driven by a *magnetic motive force* mmf, analogous to the emf of an electrical circuit. mmf may be due to a permanent magnet or a current-carrying wire. The field strength is proportional to the product of mmf and the *magnetic susceptibility* μ of the medium. Thus the field strength of a current-carrying coil is increased when iron or any other high-μ material is inserted into its core.

14.5   Electromagnetism

Magnetism and electricity seemed to be completely independent phenomena, until the early 1800s, when Hans Christian Oersted discovered that an electric current could cause a nearby compass needle to deflect. But there were earlier indications of a connection. For example, *in 1681, a ship bound for Boston was struck by lightning. Astronomical observations then showed that the compasses were changed; the North point was turned clear South. The ship was steered to Boston with the compass reversed.*[a]

Oersted described his discovery in a memoir, published on July 21, 1820, in Latin! It drew international attention and stimulated feverish activity. Less than two months later, Dominique Arago, reporting on his confirming experiments at a meeting of the French Academy of Sciences, described how a current acts like an ordinary magnet in attracting iron filings and an ability to induce permanent magnetism in iron needles. In November of the same year, Arago and a colleague, André Marie Ampere, showed that a coil of wire produced a stronger field than a single loop.

Ampere derived a fundamental relation for the magnetic field produced by a current. Applied to a long straight wire, *Ampere's Law* states that the

---

[a]F. Cajori, *A History of Physics*, Dover, NY 1962, p. 102.

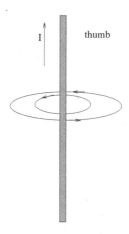

Fig. 14.3 The magnetic field of a straight wire carrying current I. The direction is given by the right hand rule.

field B at a distance r is

$$B = 2k'I/r \quad \text{T} \tag{14.4}$$

where T stands for *Tesla*, and $1\text{T} = 1 \text{ N/A-m}$, and where the constant $k' = 1 \times 10^{-7} \text{ N/A}^2$. The **B** field lines are transverse to the current, forming circles around the wire. The direction of **B** is given by a *right-hand rule*: placing your right hand so that your thumb points in the direction of the current, your fingers coil around the wire, pointing in the direction of the field. See Fig. 14.3.

A circular loop current of radius r produces a field at its center:

$$B = 2k'I/r, \tag{14.5}$$

like that of a straight wire, but r has a different meaning here. See Fig. 14.4. The field due to a coil of wire is a sum of the fields produced by each loop; if the coil has n turns, the field at its center is n times the magnitude of field produced by a single loop.

$$B = 2k'nI/r. \tag{14.6}$$

The field pattern of a loop current is the same as that of a bar magnet at the center of the loop, oriented normal to the plane of the loop. The magnetic dipole moment of the loop is the product of the current I and area S of the loop:

$$\mu = IS \quad \text{Am}^2, \tag{14.7}$$

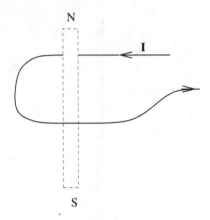

Fig. 14.4   The magnetic field of a current loop is like that of a magnetic dipole in the center of the loop.

and is directed normal to the plane of the loop, as given by the right hand rule.

Illustrative Example 14.2: What is the radius of a loop with a current of 2.0 A that would have the same magnetic moment as the bar magnet of Illustrative Example 14.1 (the bar magnet is located at the loop's center)? What is the magnetic field at the center?

Answer: $\mu = IS = 2A\pi r^2 = 1.2 \times 10^{-2}$ J/T. $r = 4.4$ cm.
$B = 2k'I/r = 9.1 \times 10^{-6}$ T.

14.6   Forces and Torques on Currents

The field due to a current interacts with any external field. A straight wire carrying a current I in a field B experiences a force per unit length

$$F/\ell = BI\sin\theta, \quad \vec{F}/\ell = \mathbf{I} \times \mathbf{B} \tag{14.8}$$

where θ is the angle between the directions of I and B. Accompanying this force is a reaction on the source of B, such as the compass needle in Oersted's famous experiment.

A current-carrying loop experiences a torque in an external magnetic field, tending to rotate the loop until its plane is perpendicular to the field. This torque is the basis for electric motors, and for the *galvanometer*,

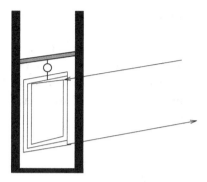

Fig. 14.5 The basic construction of a d'Arsonval galvanometer, which is the basis of both ammeters and voltmeters.

which is the mechanism for current and voltage measuring instruments. See Fig. 14.5 (see Sec. 4.10).

Currents exert forces on each other due to the interaction between their magnetic fields. Two parallel wires a distance r apart, carrying currents I_1, I_2 exert a force/unit length

$$F/\ell = 2k'I_1I_2/r\,. \tag{14.9}$$

The force is attractive if the currents are flowing in the same direction; repulsive if in opposite directions. You can see why if you examine the magnetic fields of the two currents in the space between the wires. They are in opposite directions. When the currents are in opposite directions, the magnetic fields between the wires are in the same direction. The attractive interaction ("pinch effect") between parallel currents in straight wires has been used in an experimental project that is attempting to produce a practical energy source by nuclear fusion.

14.6.1 *Electric meters and motors*

A *galvanometer* is a current-metering instrument based on the force exerted by the interaction between a current and a magnetic field. A simple design has a coil between the poles of a permanent magnet. See Fig. 14.5. A current produces a field through the coil that tends to rotate it; and the rotation is restrained by a spring. Thus the coil rotates to an angle that increases with the current. The attractive interaction ("pinch effect") between parallel currents in straight wires has been used in an experimental project that is attempting to produce a practical energy source by nuclear fusion.

There are various designs of motors, but all are based on the force between a current carrying circuit and a magnetic field, which may be produced by a permanent magnet or by field coils. The rotary member is a coil mounted on a shaft, and connected to an emf via the sliding contact of *brushes* or *slip rings*.

14.7   Force on a Charged Particle

Magnetic fields act on moving charges, whether in a wire or not; all charged particles experience forces when they move through a field. A particle of charge q traveling at velocity $\mathbf{v}$ through a field $\mathbf{B}$ experiences a force (named the *Lorentz force*) described by the equation (see Fig. 14.6)

$$\mathbf{F} = q\mathbf{v} \times \mathbf{B}. \tag{14.10}$$

The speed v is constant if B is constant. The Lorentz force acts at a right angle to the particle's velocity; thus, it tends to force the particle to move in a circle. We derive the circle's radius R by equating the Lorentz force to the inertial force causing a centripetal acceleration. Thus, $qvB = mv^2/R$, so that

$$R = mv/qB, \tag{14.11}$$

The Lorentz force due to the Earth's magnetism provides a zone of protection against high energy charged particles from *cosmic rays* and the

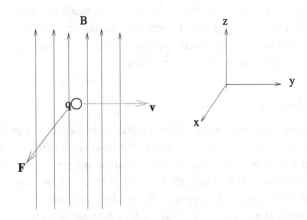

Fig. 14.6   The Lorentz force on a charged particle moving in a magnetic field.

Sun. The particles are deflected by the field extending over most of the Earth's latitudes, except for directions near the polar axis. Charged particles bombard the atmosphere in the Polar regions, where they ionize the upper regions. They are responsible for visual displays, known as the *Aurora Borealis* in the North, (also called *the Northern Lights*), and the *Aurora Australis* in the South.

The Lorentz force is put to use in particle accelerators. In the *cyclotron* and other high energy accelerators, a strong magnetic field keeps charged particles circulating in confined orbits, while periodic changes in electric fields pump energy into the beam.

14.8 Earth's Magnetism

The source of the Earth's magnetism is a dynamical phenomenon, known as the *magneto* effect. It involves the flow of electrically conducting lava in the Earth's core. The Earth's high internal temperature is due to a combination of the heat due to radioactivity and the internal friction due to tidal motion. The dense liquid, thought to be mainly a mixture of iron and nickel, flows slowly in convection patterns driven by the heat escaping outward toward the surface and the Earth's rotation. Being electrically conducting, the convecting liquid carries electrical currents, creating a magnetic field, which resembles the field around a bar magnet. Magnetic fields are detected around some planets; interpreting these fields as due to the same mechanism as the Earth's magnetism, it is deduced that their cores are similarly composed of molten metals.

The convective flows in the core of the Earth are the driving forces for *continental drift*. The continents have been moving, as islands afloat on the surface of the planet, during much of the Earth's history. One type of evidence is in the congruent shapes of some continental boundaries; e.g. South America can be fitted neatly against West Africa. Another clue is the similarities in the geology of some areas of separate continents, showing that at some time in the past they had been joined together. And still another involves the geology of the sea bed.

There are great seams along the sea floor, where the rocks belonging to two adjacent continents are gradually pulling apart, and fresh *magma* is flowing in. This phenomenon, *sea floor spreading*, is very slow, on the order of millimeters or centimeters per century, but over the course of the Earth's history, on the order of 2 billion years, the movement can amount to a major fraction of the Earth's circumference.

During its long history, the Earth's magnetic field has changed its direction drastically. The changes are recorded in the sea bed by bands of magnetization, pointing in different directions, near the seams. The magnetization induced by the Earth's magnetic field is "frozen" into the lava as it flows up and then cools below the ferromagnetic Curie temperature. Thus we see that the field has reversed several times; what is now the South magnetic pole has been the North magnetic pole, and vice versa. The cause of the reversals is not known.

Questions

Q14.1. Why is the magnetic field of a coil stronger if it has an iron core?

Q14.2. What is the cause of the Aurora Borealis?

Q14.3. Why does continental drift suggest that the continents have lower density than the deeper rock?

Q14.4. Describe how a magnetic field can "trap" charged particles.

Q14.5. What is the direction of the force on a positively charged particle moving S to N next to a wire carrying a current from N to S?

Q14.6. A horizontal wire carries a current from West to East. In what direction would a compass needle point if placed below the wire?

Q14.7. A circular wire has a current moving counterclockwise. What is the direction of the magnetic field in the center? Where would the N pole be?

Fig. 14.7   A rectangular loop of current-carrying wire.

Q14.8. For the current shown in Fig. 14.7, find the direction of the magnetic field for each segment of the current-carrying wire.

Q14.9. Why does the N pole of a compass needle point N and not S?

Q14.10. Two identical long bar magnets have S poles separated by 3 cm. If the separation is doubled, by how much does the force of repulsion between the S poles change?

Problems

P14.1. How much work is done by **B**, when it causes a particle of charge q and velocity **v** to move in a circle?

P14.2. Two 15.0 cm long long bar magnets are lying on a table, with their N poles facing each other. The N poles are 2.0 cm apart, and exert a 2.0 N force of repulsion on each other.
(a) What is the force between the S poles?
(b) What is the total force between the unlike poles?

P14.3. A magnetic field of magnitude 2.0 T points toward the East. An electron traveling at 2.5×10^6 m/s enters the field.
(a) In what direction does the electron travel, if it feels zero Lorentz force?
(b) The electron's direction is N; calculate the magnitude and direction of the Lorentz force.

P14.4. A bar magnet with a dipole moment $\mu = 1.0 \times 10^{-4}$ J/T is placed in a magnetic field of 0.1 T that points North.
(a) The magnet points to the West. Calculate the magnitude and direction of the torque.
(b) What is the magnet's orientation energy when it aligns parallel to the field?

P14.5. Two long parallel wires, each carrying a current of 5 A, are 10 cm apart. The currents are in opposite directions.
(a) Is the force between the wires attractive or repulsive?
(b) How large is the force/unit length?

P14.6. An electron is traveling at 5.0×10^4 m/s in a 1.0 T magnetic field that points Eastward.

(a) What is the electron's direction if there is no magnetic deflection?

(b) What is the force if it is traveling towards the S?

P14.7. The magnetic force on a 20.0 cm length of straight wire in a perpendicular 2.0 T magnetic field is 1.0×10^{-4} N. Calculate the current through the wire.

P14.8. What is the orbital radius of a 10.0 keV electron (1 eV = 1.6×10^{-19} J) in a 2.0 Tesla field?

P14.9. Two parallel straight wires 15.0 cm apart are carrying currents of 3.0 A in opposite directions.

(a) Is the force between them attractive or repulsive?

(b) What is the force/unit length between them?

P14.10. A wire loop has a radius of 0.10 m, and carries a current of 0.50 A. It is oriented at 45° to a 0.2 T magnetic field.

(a) How large is the torque on the circuit?

(b) How large is its orientation energy?

P14.11. A 50-turn coil, which has a resistance of 150 ohms, is wound on a circular frame that has a radius of 0.20 m. The coil is connected to a 9 volt storage battery. Calculate the magnitude of the magnetic field in the center of the coil.

P14.12. A 150 MeV electron (1 MeV = 1.6×10^{-13} J) traveling from S to N encounters a constant magnetic field of 2.5 T oriented E to W. Show the path of the electron and determine the radius of its path.

Answers to odd-numbered questions

Q1. The high magnetic susceptibility of the iron increases the magnetic field in the circuit.

Q3. Continental drift shows that the continents float on the deeper rock of the Earth's "mantle".

Q5. Repulsive.

Q7. If the paper is the plane of the circle, then $\vec{B}$ is upward. The N pole would be up from the paper.

Q9. Because the N magnetic pole is actually a S pole of a bar magnet in the earth.

Answers to odd-numbered problems

P1. The field does no work, because the magnetic force is perpendicular to the motion of the particle.

P3. (a) If the electron travels parallel to the field, it feels zero force.
(b) 8.0×10^{-13} N, directed perpendicular to the $\mathbf{v}, \mathbf{B}$ plane.

P5. (a) The force is repulsive.   (b) 5×10^{-5} N/m.

P7. 2.5×10^{-4} A.

P9. (a) Repulsive.   (b) 1.2×10^{-5} N.

P11. $3.0 \ \mu$T.

Chapter 15

Electromagnetism — II

15.1 Michael Faraday

Michael Faraday (1791–1867) was one of the most creative scientists of the modern era. He has been called "The father of the electric age", yet he had a most rudimentary education, leaving school after the 4th grade. He was taught no mathematics beyond arithmetic. As a teenager he apprenticed to a bookbinder, and then began educating himself by reading the books that came in for binding. One day a customer offered him a ticket to a public lecture to be given by Sir Humphrey Davy, who was one of the eminent scientists of the age. Faraday was captivated; he took careful notes and sketches, bound them into a book as a gift for Davy, who was so impressed by the book and the boy that he hired him as a laboratory assistant. That began his life of science, in the course of which he made pioneering discoveries in chemistry and physics, published over 200 papers, and was elected to every scientific academy in Europe. See Fig. 15.1 for a picture.

15.2 Generating Electricity from Motion

Our modern technological society would not exist if we had no way of generating electricity, other than by triboelectricity and batteries; no mechanism for converting mechanical energy into electricity. But we can do just that, by Faraday's discovery of an effect known as *electromagnetic induction.*

In his research toward induction, Faraday's thought processes may have gone something like this:

Michael Faraday

Portrait of Michael Faraday by Thomas Phillips
(1841–1842)

Fig. 15.1 Portrait of Michael Faraday, courtesy of the National Portrait Gallery, London.

If a magnetic field tends to produce motion of a wire carrying an electric current, can the reverse also work? That is, can the motion of a wire in a magnetic field produce a current?

The thought was followed by a series of experiments, which he described in great detail in his journal.[a] He eventually succeeded with a simple arrangement — a coil of wire and a magnet. Moving the magnet into or out of the coil, or just moving it near the coil, induced a momentary current in the coil.

When a politician asked him, "What possible benefit could come from his idea?" Faraday answered that he didn't know, but he was sure that if it was put to a practical use, politicians would tax it!

A wire loop of area A is in a magnetic field $\mathbf{B}$. There is *magnetic flux* Φ through the loop, defined as the product of the area and the component

---

[a]Faraday's Diary, G. Bell and Sons, London 1932, 1936.

of **B** that is perpendicular to the plane of the loop:

$$\Phi = B_\perp A = BA \cos\theta = \vec{B} \cdot \hat{n} A, \qquad (15.1)$$

where θ is the angle between **B** and the unit vector $\hat{n}$, normal to the plane, $\hat{n}$. The MKS unit of Φ is $T - m^2$.

Faraday's marvelous discovery is now known as *Faraday's Law: A flux change induces a momentary voltage (emf) while the flux is changing*:

$$\mathcal{E} = -\Delta\Phi/\Delta t, \qquad (15.2)$$

where Δt is the time duration of the change.

For a coil of wire, with n turns of area A, the induced emf is n times the emf of a single loop. The equation holds true for all flux changes, whatever their origins, including relative translational motion of a magnet and coil, rotation of the coil, or changing strength of the magnetic field. If the flux change is measured in Tesla $- m^2/\text{sec}$, the emf is in *volts*. The direction of the emf is such as to induce a current that opposes whatever created it. If the rate of flux is increasing in a coil of wire in the plane of the paper, and the magnetic field is out of the paper, then the emf will drive a current clockwise around the loop, so as to produce a magnetic field into the paper, opposite to the increase of the original field (out of the paper). It is in opposition to the flux change.

The negative sign means that *any induced current from the emf is such as to oppose the action which causes it.* For instance, if the flux is decreasing, the current in a circuit connected to the induced emf will oppose the reduction — it will generate a magnetic field that would increase the flux. The induced emf is generated whether there is a conducting path or not. As we will see in the next chapter, this relation is an element of the physics of electromagnetic waves.

15.2.1 *Lenz's law and Eddy currents*

When current is induced by the relative motion of a circuit and a magnetic field, the current creates a magnetic field that opposes the motion. This principle was enunciated by a Russian physicist, Heinrich Lenz (1804–1865), and is the reason for the minus sign in Eq. (15.2). Lenz's Law assures that the induction requires work by an external agency. An example of the principle is exhibited by "eddy current drag". Induction of a current in a conducting circuit requires work by an external agent to create the electrical energy. Eddy current drag is the basis for a number of devices, such as speedometers in cars, and a braking system in electric buses and trolleys.

15.2.2 *Electrical power generators*

Equation (15.2) is the basic relation that governs the generation of electricity in electric power plants. The mechanical motion may be produced by any one of a number of methods; heat engine, wind power, water power, or nuclear energy, but the conversion to electricity always involves the physics of Eq. (15.2).

In typical generators the coil rotates at a steady rate, with angular velocity $\Delta\theta/\Delta t = \omega$. Then, one can show graphically (or by differential calculus), that Eq. (15.2) for an n-turn coil leads to a periodically changing emf:

$$\mathcal{E} = -nBA\omega \cos(\omega t) \quad \text{or} \quad \mathcal{E} = -nBA\omega \sin\omega t, \qquad (15.3)$$

depending on whether the loop starts parallel or perpendicular to $\vec{B}$, respectively.

Thus, the generator produces an *alternating voltage* cycling from positive to negative with frequency $f = \omega/2\pi$. The standard "household power" in the US and most countries with a national power system has a root-mean-square emf of approximately 115 volts, and angular velocity of $\omega = 2\pi \times 60$ radians/s, corresponding to a frequency $f = 60$ Hz.

Illustrative Example 15.1: A circular coil of wire of radius 3 cm is in a magnetic field $\vec{B} = 0.2$ T oriented at 60° relative to the plane of the loop.
(a) What is the flux threading the loop?
(b) If the magnetic field is turned off in a time of 0.2 s, what is the average induced emf in magnitude and direction. Show graphically.

Solution: (a) The flux is $BA\cos\theta$, where θ is the angle relative to the normal of the loop $= 30°$. $\Phi = 0.2$ T$(\pi r^2)\cos(30°) = 4.9$ T $-$ cm$^2 = 4.9 \times 10^{-4}$ T-m^2.
(b) $\mathcal{E} = +4.9$ T-cm$^2/0.2$ s $\approx 2.5(10^{-3})$ V. If the coil is in the plane of the paper and the magnetic field that is being turned off is upward, then the induced emf would be such as to generate an upward magnetic field. The current in the loop is counterclockwise. This generates a magnetic field directed upward, out of the paper, in accord with Lenz's law.

There are many other applications of Faraday's discovery. Electric guitars use a set of vibrating strings made of magnetizable metal. The pick-up is a coil of wire with permanent magnetism. When you pluck a string on the guitar, the flux changes in the coil and an emf is produced e.g. at 440 Hz. This emf or the resulting current can be amplified.

Microphones can have a coil attached to the vibrating diaphragm; if it is placed near a stationary magnet, the motion of the diaphragm changes the flux and induces an emf. The signal can be amplified.

15.3 Transformers

Electrical transformers can increase (*step up*) or decrease (*step down*) the voltage of an AC source. An AC source is connected to the transformer's *primary coil*, which produces an AC magnetic flux proportional to the current in the primary coil: $B \propto I_p$. The current is proportional to the voltage $\mathcal{E}_p$ and inversely to the number of turns: $B \propto \mathcal{E}_p/N_p$. A secondary coil is coupled to the primary coil, either by winding it tightly around it, or by an iron core that is shared by both. The flux though the secondary coil is proportional to the field and the number of turns of the secondary: $\Phi \propto N_s \Delta B/\Delta t$. Thus

$$\mathcal{E}_s/\mathcal{E}_p = N_s/N_p. \tag{15.4}$$

Illustrative Problem 15.2: If you need 6 V and only have house voltage of 115 V, what is the number of secondary coil turns you have to use in a transformer with 100 turns in the primary?

Solution: $\mathcal{E}_s = (N_s/N_p)\mathcal{E}_p$, or $N_s = N_p(\mathcal{E}_s/\mathcal{E}_p) = 100(6/115) = 5.22$ or approximately 5 turns.

15.4 Battle over Electric Power

The development of "standard" electric power in the U.S. has a history of conflict. Thomas Alva Edison (1897–1931) was a self-educated man, an enormously energetic inventor, and a self-publicist. He built a research laboratory in New Jersey that later merged with another firm to become the General Electric Company. In the course of his career he obtained patents for the incandescent light bulb, dictaphone, mimeograph, storage battery, sound movies, and the phonograph.

In the early 1920's, he planned to 'electrify' New York City. At that time, the principal mode of street lighting was the gaslight. Edison's idea was to install electric generators throughout the city. Because the electrical resistance of long wires would waste power, the generators would be closely spaced so that the transmission distance would be short. Edison based his plan on direct current (DC) power. (DC power is obtained from AC generators by equipping them with *commutators* — split rings which transform AC to DC).

Nikola Tesla, who was then a young engineer who had worked at Edison's company, proposed a different idea. With alternating currents and *power transformers*, it would be possible to transmit power over long distances with low resistive losses. The voltage output of the power plant would be "stepped up" to a high voltage by a transformer for the transmission from the power plant to its destination. The resistive losses are much smaller at the higher voltage, because the current is much lower (VI is constant for a transformer). The voltage is stepped down again for safety at the end point, a local group of houses, factories, or street lights. Thus, one power plant could supply electricity efficiently to a wide area. The AC system won out because of its efficiency and consequent financial advantage.

Tesla and Edison waged a publicity battle. Edison claimed that AC power was more dangerous because it involved peak voltages that are much greater than the average (although they are only about 40% greater). He demonstrated this by electrocuting puppies, a horse, and an elephant with AC voltage. (He made a movie of their electrocution).

Questions

Q15.1. Why don't birds get electrocuted when they land on high voltage transmission lines?

Q15.2. A rectangular loop of wire in the plane of the paper, is moved so that more and more of it is in a well-defined magnetic field. If the magnetic field is into the paper, what is the direction of current flow in the loop? Why?

Q15.3. It has been proposed that electric power can be generated by the rotation of the Earth, if long wires are installed at right angle to the Earth's magnetic field. Can it work?

Q15.4. A conducting plate is pivoted so that it can swing freely in its plane, as a pendulum. A magnet is brought close to the plate, so that the plate is suspended between the pole pieces, and the motion is perpendicular to the field. What is the magnet's effect on the oscillations of the plate? Compare the effects for plates that are composed of pure copper (low resistivity) and stainless steel (high resistivity). Think of the bob of a pendulum oscillating in a magnetic field. (*Hint:* The current would go in circles around the bob.

Q15.5. A rectangular loop of wire is in the plane of the paper. A magnetic field $\vec{B}$ is going South to North in the same plane.
(a) If there is a clockwise current in the loop, what is the direction of the torque?
(b) If the loop turns around an axis perpendicular to $\vec{B}$ (from E to W) in the plane of the paper and centered on the loop, is the induced emf a maximum or zero at the start? Explain.

Q15.6. Repeat Q15.5 if the loop is in the plane of the paper and the magnetic field is out of the paper. Explain.

Q15.7. Coil A and coil B have the same radius, but coil A has twice as many turns as coil B. If a magnetic field perpendicular to the area of the coils increases at the same rate for both coils, what is the ratio of the induced voltages (emf's)?

Q15.8. (a) For the loop of Q15.6 (with no rotation of the loop), if the magnetic field decreases with time, is the current induced clockwise or counterclockwise? Why?

Q15.9. Can you use a transformer with a DC current in the primary? Explain.

Problems

P15.1. A magnetic field that passes perpendicularly through a circular loop of wire changes from 0.03 T to zero in 5 seconds. The radius of the loop is 50 cm. What is the average emf generated in the coil?

P15.2. Ten turns of wire are wound on a wooden frame that measures 10.0 cm × 20.0 cm. The coil is placed in a 0.10 Tesla magnetic field, with the plane of the coil perpendicular to the field.

(a) How much flux is enclosed by the coil?

(b) The plane of the coil is turned 90 degrees in 5 seconds. What is the average emf during the motion?

(c) Is the emf a maximum or zero at the start? Explain.

P15.3. A loop of wire enclosing an area of 0.010 m² is in a 0.50 T magnetic field, with the plane of the loop inclined at 45 degrees from the direction of the field. What is the flux enclosed by the loop?

P15.4. A wire loop that measures 20.0 cm × 30.0 cm is placed in a 0.020 T field from a permanent magnet, with the plane of the loop perpendicular to the field.

(a) How much flux is enclosed by the loop?

(b) The magnet is removed from the neighborhood of the coil in 5 seconds. What is the average emf generated in the loop?

P15.5. A transformer has a primary coil of 500 turns and a secondary coil of 100 turns. The primary is connected to a 110 volt power source. What is the voltage output of the secondary coil?

P15.6. A circuit consists of two conducting rails 1.2 m apart and 20 m long. At one end of the rails is a permanent cross bar. At the other end of the rails is a sliding, conducting cross bar that moves so as to decrease the length of the rails within the loop. The "circuit" is placed in a magnetic field, $\vec{B}$ perpendicular to the plane of the rails. The resistance of the circuit is an average of 3.2 ohms and the change of R during the sliding of the bar can be neglected.

(a) Find an expression for the induced emf if the cross bar moves at a velocity v.

(b) Find the value of the induced emf if the cross bar moves at 1.2 m/s.

(c) Determine the direction and magnitude of the induced current if the rails are in the plane of the paper and $\vec{B}$ is out of the paper.

P15.7. A power generator has a rotating coil of 500 turns, and its area encloses a flux of 0.01 T-m² for each turn. The coil turns at the rate of 60 revolutions/second.

(a) What is the frequency of the induced voltage?

(b) What is the peak voltage?

(c) Does the peak voltage occur when the coil is parallel or perpendicular to the magnetic field?

P15.8. A transmission line that has a resistance of 20.0 ohms connects a power generator to a factory that is consuming power at the rate of 1500 kW. The output voltage of the generator is 50 kV.

Hint: The power loss is not V^2/R because the voltage is *between the conductors*.

(a) What is the power loss due to the line's resistance?

(b) The power station steps up its output voltage to 500 kV by a transformer. The factory steps the voltage back down to 50 kV. The power station delivers the same power to the factory as it had before the conversion. What is the resistive loss in the transmission line when it is operated at 500 kV?

Answers to odd-numbered questions

Q1. Because there is only a small difference of potential between the two feet.

Q3. No. There is no relative motion between the wires and the field.

Q5. (a) The torque is from W to E.
(b) The flux is zero, but the change of flux is a maximum.

Q7. The induced voltage is proportional to n.

Q9. No — you need a change of flux.

Answers to odd-numbered problems

P1. 4.7×10^{-3} V.

P3. 3.5×10^{-3} T-m^2.

P5. 22 Volts.

P7. (a) 60 Hz.
(b) $500 \times 0.1 \times 60 \times 2\pi = 18.8$ kV.
(c) When the coil is parallel to the field, where the change of flux is highest.

Chapter 16

Electromagnetic Waves

16.1 Electromagnetic Waves

We studied waves in Chapter 9. What do radio waves, light, and X-rays have in common? They are all electromagnetic waves. *Electromagnetic waves* are transverse waves that do not require a medium for propagation. Light is an example of an electromagnetic wave and traverses vacuum between the sun or the stars and us. Until the beginning of the 20th century, it was believed that light required a medium, dubbed *aether* and that this substance permeated all space. Einstein showed that it was unnecessary and we shall see later how its non-existence was demonstrated. There are many different kinds of electromagnetic waves. Examples are radio waves, microwaves, infra-red, ultra-violet, X-rays and gamma rays, in order of decreasing wavelengths. Different forms of electromagnetic waves are produced by different means. The spectrum of electromagnetic waves is shown in Fig. 16.1, which also gives the range of frequencies and wavelength for each kind of wave. The range of visible light wavelengths is roughly 3.8×10^{-7} m (violet) to 7.5×10^{-7} m (red). As for all waves, the wavelengths, λ and frequencies, f are related to the velocity of wave propagation by

$$v = f\lambda. \tag{16.1}$$

Electromagnetic waves were predicted by James Clerk Maxwell in 1865, who also formulated the basic laws of electromagnetism. He predicted the velocity of all electromagnetic waves in vacuum to be equal to that of light, $c = 3.00 \times 10^8$ m/s.

How can you measure the speed of light? Galileo gave it a try with an assistant and lanterns that could be opened and shut. One of the men was

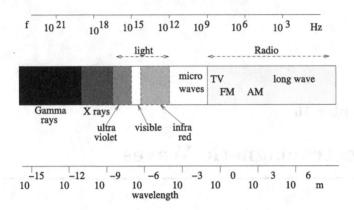

Fig. 16.1   The spectrum of electromagnetic waves.

on a small hill and the other one on a distant hill. When one opened the lantern, the other responded by opening his, but light was too quick for them and they did not succeed. The response time of the individual is too slow. It only takes about 1 s for light from the earth to reach the moon! On land, success was achieved by Armand-Hyppolite Fizeau in 1849. He used a toothed wheel to shine light through and placed a mirror almost 9 km away. If the wheel rotates sufficiently fast, then it will be at the succeeding notch when the light returns and can be seen by the observer. The apparatus is sketched in Fig. 16.2.

Later, Albert Michelson used an octagonal rotating mirror instead of a notched wheel to remeasure the speed. In 1675, Ole Roemer, a Danish physicist used the times between eclipses of Io, one of the satellites of Jupiter to measure the velocity of light. He found that the time is longer when the earth moved away from Jupiter than when it moved towards it. He reasoned that this delay must come from the finite speed of light. At the time, the distances he needed were not known, so Roemer was unable to deduce a value for c.

When we examined electric and magnetic fields, they were only concepts, which made forces acting at a distance easier to understand. With waves, these fields acquire a reality. It is the electric and magnetic fields which do the vibrating in electromagnetic waves. This explains why no medium is necessary for their propagation.

Consider the dipole antenna in Fig. 16.3. It is connected to an a.c. generator. When the charges in the upper arm are + and those in the lower one −, the electric field is downwards. One half cycle later, the situation is

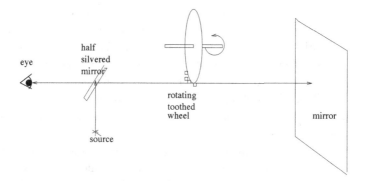

Fig. 16.2   Sketch of the apparatus used by Fizeau to measure the speed of light.

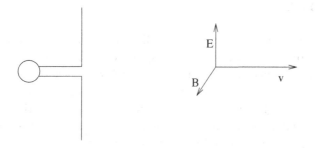

Fig. 16.3   A simple radio (dipole) antenna. Relation of $\vec{E}$, $\vec{B}$, and $\vec{v}$.

reversed. The motion of the charges gives rise to a current, which generates a magnetic field at right angles to the electric one. A changing magnetic field gives rise to an electric one, and a changing electric field gives rise to a magnetic one. As the fields move away from the antenna, they generate electromagnetic waves; these waves are periodic since the current that generated them were periodic. Radio waves are generated in this manner.

Although Maxwell predicted electromagnetic waves, Heinrich Hertz was the first to demonstrate their propagation in 1888. He used a circular loop, but later also did experiments with straight wires. He detected the waves with another circuit at some distance from the source antenna.

Radio waves have many uses beyond providing entertainment, e.g. communication, ranging, GPS, radar. They are also used to detect distant stars by means of large curved (dish-like) antennas to focus the radiation onto a receiver. The time for light or radio waves to reach the earth from distant stars is long. The waves that reach us today from large distances were emitted many years ago. This allows us to peer back in time and observe conditions in the early universe. In 1987, a rare explosion allowed us to witness the death of a star (supernova 1987 A) in a neighboring galaxy, about 1.7×10^{21} m away. The time to get here from this distance is 1.7×10^{21} m$/(3 \times 10^8$ m/s$) \simeq 5.5. \times 10^{12}$ s or approximately 135,000 y.

Infrared electromagnetic radiation is used in microwave ovens to deliver energy to food. Infrared electromagnetic waves also play a role in the greenhouse effect. The presence of CO_2 and CH_4 in the atmosphere blocks infrared radiation from the sun to the earth and reflects it back into space. They also reflect escaping heat from the earth back to the earth.

16.2 Light

Light is undoubtedly the most familiar electromagnetic wave to us, since it is all around us and we can hardly do without it. For the early Greeks, light was believed to be composed of particles which hit the eye. Newton also believed that light was made of particles. It was not until the 19th century that light was shown to be a wave. A characteristic of a wave that we have already seen is interference (superposition of waves). We shall see other ones: refraction and diffraction — i.e., the bending of light around obstacles and corners. Later in the 20th century, it was shown that light can behave as either a wave or as particles, but for the time being, we shall study the wave properties of light.

Why was it so difficult to show that light is a wave? It is because the wavelength of light, unlike sound, is very small, of the order of 10^{-7} m. To us, it appears that light travels in straight lines; that is why we see shadows in sunlight and why pinhole cameras work. To see the wave properties you need objects of the order of the wavelength, and these are not readily available.

In Fig. 16.4, we show *rays* of light and *wavefronts* from a "point" source. The rays are light beams emanating from the source, whereas wavefronts are perpendicular to rays and represent a fixed phase of a wave, such as the crest. The wavefronts are spherical surfaces. For a source that is very far away, these surfaces become planes at right angle to the rays. In the 17th century, Christian Huygens showed that each point of a wavefront can be considered as the source of new wavelets. The tiny wavelets form a new wave. This principle is useful in refraction, and we will see it again later.

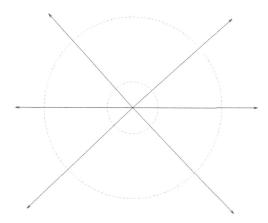

Fig. 16.4   Rays and wavefronts from a point source of light.

16.3   Refraction

Refraction is the bending of light when it enters a different medium. The frequency of the wave cannot change at the surface or in the new medium. But the velocity of light depends on the medium it travels in, and, thus the wavelength of light is altered. The decrease in velocity in a medium is due to the excitation and coherent reradiation of molecules in the medium. For most media, the velocity of light is decreased relative to that in vacuum,

$$v = c/n \,, \tag{16.2}$$

where n is called the *index of refraction* of the medium. Generally, the index of refraction is larger than 1. It is approximately $= 1$ for air at normal pressure and temperature. The decrease in velocity of light causes a proportional decrease in the wavelength and for any two media, we have

$$\frac{\lambda_2}{\lambda_1} = \frac{v_2}{v_1} = \frac{n_1}{n_2} \,, \tag{16.3}$$

When light enters a medium with a larger index of refraction than the one it came from, it is bent towards the normal (a line perpendicular to the surface). The reverse is true when the situation is reversed,

$$\frac{\sin \theta_2}{\sin \theta_1} = \frac{n_1}{n_2} = \frac{v_2}{v_1} \,, \tag{16.4}$$

where the angle θ is measured from the normal to the surface, see Fig. 16.5. This relationship is known as *Snell's law*.

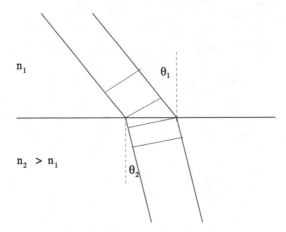

Fig. 16.5   Refraction of light entering a different medium.

The index of refraction of water is approximately 1.33 and that of normal plate-glass is 1.5, but it can be larger in special glasses such as leaded glass.

Refraction of light can be understood geometrically by drawing wavefronts, as in Fig. 16.4, where they are shown for a distant source. Since the velocity and wavelength of light is smaller in a denser medium, the wavefronts (e.g., maxima of the wave) are closer together. This causes the wave and the rays to bend towards the normal, as shown in Fig. 16.5.

There are many distortions which occur due to the changes in the velocity of light at a surface with another medium. A pencil in water appears to have a kink when viewed from above. A fish appears to be closer to the surface than it actually is, as shown in Fig. 16.6. By contrast, to the fish you appear further away than you actually are. If the ray angles are small, you can show that the distance the fish appears to be under water, h' is

$$h' = h\frac{n_{air}}{n_{water}}, \tag{16.5}$$

where h is the actual distance the fish is under water. We have

$$\frac{\tan\theta_R}{\tan\theta_i} = \frac{x/h}{x/h'} = h'/h \approx \frac{\sin\theta_R}{\sin\theta_i} = \frac{n_i}{n_R}, \tag{16.6}$$

$$h' = h\frac{n_i}{n_R} \tag{16.7}$$

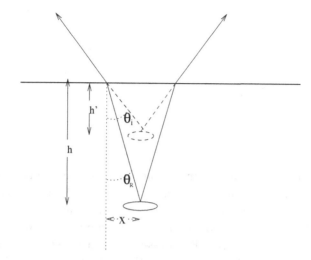

Fig. 16.6 A fish in water and rays emanating from it. The dashed lines shown where the fish appears to be located.

Illustrative Problem 16.1: Red light of frequency 4.76×10^{14} Hz is incident on a special glass plate of index $n = 1.44$ at an angle of $60°$ relative to the plane of the glass.
(a) What is the angle of the refracted ray?
(b) What is the velocity of the light in the glass?

Solution: (a) $\sin(\theta_{refr}) = \sin(30°)/1.44 = 0.347$. $\theta_{refr} = 20.3°$.
(b) $v_{glass} = 3 \times 10^8/1.44 = 2.08 \times 10^8$ m/s.

Illustrative Problem 16.2: A fish in the water ($n = 1.33$) sees you and your eyes at 2.1 m above the water surface. What is the real height of your eyes?

Solution: $h' = h(n_R/n_i)$; $h = h'(n_i/n_R) = 2.1/1.33 = 1.58$ m.

We can understand the colors of the rainbow in the same manner. Light from the sun is bent at the surfaces of raindrops. If the sun is behind you the outgoing refracted light is dispersed. As shown in Fig. 16.7, the top of the rainbow will be red and the bottom blue or purple, since red light is deflected least and purple most. A secondary rainbow can occur if there is more than one reflection; see Fig. 16.7.

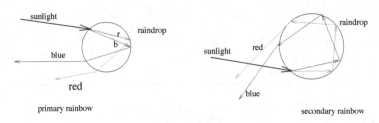

Fig. 16.7   The formation of rainbows.

When light is incident from a medium of higher index of refraction to one of lower index, it is bent away from the normal. What happens when the angle in the less dense medium reaches 90°? For any angle larger than this *critical* one the light cannot get out of the medium with larger index and is *totally internally reflected*. For glass-air interface, *total internal reflection* occurs when the angle of incidence in the glass, with $n = 1.5$, (with respect to the normal), is

$$\frac{\sin \theta_{glass}}{\sin 90°} \approx \frac{1}{1.5}, \tag{16.8}$$

$$\theta_{critical}(glass) \approx 41.8°. \tag{16.9}$$

This feature is often used in optical instruments and also helps to explain some of the brilliance of diamonds. Total internal reflection is used in light pipes and fiber optics, which use materials with a high index of refraction, so that the critical angle is small. Almost all the light is then transmitted along the pipe or fibers.

16.4   Color

We know that visible light comes in a variety of colors. The longest wavelength is red at 620–750 nm or 6200–7500 Å(1 Å$= 10^{-10}$ m). Purple has the shortest wavelength in the visible region, $\lambda = 380$–440 nm. White light and sunlight contain all the colors, as was demonstrated by Newton. If you mix red and green, it is perceived as yellow; this is called additive color mixing. You can combine the three primary colors, blue, green and red in varying amounts and get all colors including white. You can also do subtractive color mixing, e.g. by using filters to remove some colors.

The observation of color depends on the wavelengths present in the source and in the manner that the object reflects, absorbs, or scatters light.

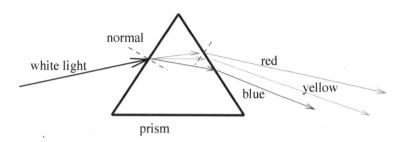

Fig. 16.8 Use of a prism to disperse colors.

Dyed wool tends to absorb certain colors; this selective removal of colors is a subtractive process and we see these substances by the colors that remain. Objects are also seen by the reflection of light, which can be *specular* (mirror-like) or diffuse. In the latter case, light is scattered in many directions due to the roughnesss of the surface or the penetration of light into the interior. For instance a solid that reflects red light more than other colors will appear to be red in sunlight. If you insert a filter that absorbs red light, the object will appear to be blue-green. When you mix colors, you get a different color.

Newton made a serious study of light and published his theory of colors in 1672 at the age of 29. His demonstration of the colors contained in white light used prisms. What causes the separation of colors? Different wavelengths are refracted through different angles, because the index of refraction increases slightly with increasing frequency. Thus, red light is bent the least and purple the most. This *disperses* the various colors in the beam as can be observed by projecting the output of the prism on a screen; see Fig. 16.8.

When Newton demonstrated his separation of color, skeptics did not believe him. It was thought that the different colors originated in the prism. Newton showed that this was not correct by using different incident colored lights.

We can also understand why the sky looks blue during a sunny day and reddish in the evening when the sun is setting. The light is scattered by air molecules, that is, it is deflected through various angles. The scattering of light depends strongly on wavelength ($\propto 1/\lambda^4$), so that blue light tends to be scattered the most. This follows because the wavelength of blue light is among the smallest and thus closest to the size of molecules ($\sim 10^{-9}$ m). During the day, it is the scattered light we see because we cannot look at the sun directly. This scattered light is mostly blue. If the air is removed, e.g., at very high altitudes, the sky looks black, except when you look at the sun or the stars. In the evening, you can almost look at the sun directly and thus see the sky as red; again the blue light is scattered out of the beam.

16.5 The Eye and Color

The eye is a complex apparatus; through its connection to the brain we interpret and differentiate colors.

Fig. 16.9   Sketch of an eye.

Light enters the eye through the cornea, iris, and lens, which focuses the light onto the retina. See Fig. 16.9. The latter is made up of cells sensitive to light, namely *rods* and *cones*, located primarily in an area called the fovea. Rods are responsible for night time and peripheral vision, whereas the cones allow us to see details and distinguish colors in the daytime. There are three types of cones: short, medium, and long. The short cones are sensitive to short wavelengths, the medium ones to medium wavelengths and the long ones to long wavelengths. The brain interprets the colors perceived. When an object reflects mostly red light, the short cones are stimulated primarily and the brain interprets the color as red.

16.6 Doppler Effect

There is a Doppler effect for light, as there is for sound. However, for light it makes no difference whether it is the observer or the source that moves.

We shall see later that this follows from the theory of relativity. If the source and observer are approaching each other, the apparent wavelength and frequency of light are

$$\lambda = \lambda_0 \sqrt{\frac{c-v}{c+v}}, \tag{16.10}$$

$$f = \frac{c}{\lambda} = f_0 \sqrt{\frac{c+v}{c-v}}. \tag{16.11}$$

Here v is the speed of the observer relative to the source or vice-versa. If the observer and source are separating, then the $+$ and $-$ signs are reversed. When the source of light is receding from the observer, the frequency is lowered or light is said to be *red-shifted*. The light from other galaxies is red-shifted, showing that the universe is expanding.

The police also use the Doppler effect for electromagnetic waves from a radar gun to catch speeders.

Questions

Q16.1. What would the sky look like on the moon? Explain.

Q16.2. If an object absorbed all colors, what would be its appearance? Why?

Q16.3. The sun is visible for a short while after it sets below the horizon. Explain. Would you expect the same to be true just prior to sunrise?

Q16.4. Why does clothing under artificial light look as though it were of a different hue (color) than in broad daylight?

Q16.5. Figure 16.10 shows a prism with two angles at 45°. What happens to light incident as shown when it hits the hypotenuse? What is the angle of refraction?

Q16.6. Different frequencies in the electromagnetic spectrum are radiated by different physical systems. In general, the shorter the wavelength the smaller the system. Can you give a reason for this occurrence?

Q16.7. A certain object is painted with a coating that reflects blue and red, but absorbs yellow and green light. What color would you observe in sunlight by reflection from this object?

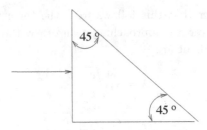

Fig. 16.10   Light entering a 45° prism.

Q16.8. How can you tell that the electromagnetic waves carry momentum and energy?

Q16.9. If the electric field of an electromagnetic wave decreases, does the magnetic field do the same, remain constant, or increase? Explain.

Q16.10. Typical X-rays have a wavelength of 2.2 nm. What is their frequency and period?

Q16.11. Prisms bend blue light more than red. Which of these colors travels faster in the glass of the prism? Explain.

Q16.12. Can an electromagnetic wave be deflected by a magnetic field? Explain.

Q16.13. Can an object have energy transferred to it by light without having linear momentum transferred? Give reason or explain.

Problems

P16.1. Astronomical distances are often given in light years, the distance that light travels in a year. How long is this unit in meters?

P16.2. One of the moons of Jupiter, Io, and its eclipse was used by Roemer to obtain the speed of light. He predicted that the eclipse in 1676 would be 10 minutes late due to the increasing distance of Jupiter from earth. He was wrong (it was more like 20 minutes), but assuming a time delay of 10 minutes, by how much did the separation of Jupiter and earth increase?

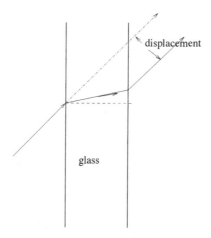

Fig. 16.11   Light going through a slab of glass.

P16.3. How long does it take light to cross a distance equivalent to the earth's diameter?

P16.4. A fish in a tank of water is examined from outside the tank. The glass of the tank enclosure has an index of refraction of 1.45 and that of the water is 1.33. The glass enclosure of the tank is 0.45 cm thick.
(a) How far behind the inner wall of the glass does the fish appear to be?
(b) Assume that this image of the fish acts as the object for the outer wall of the glass, how far behind the outer surface does the fish appear to be?

P16.5. A fish looks at your eyes, which are located 1.92 m above the edge of the water. At what vertical distance does the fish see your eyes, if the index of refraction of water is 1.33?

P16.6. A light beam traverses a slab of glass of index $n = 1.5$. The beam is displaced when it comes out, as shown in Fig. 16.11.
(a) Is the displacement larger for red or blue light? Explain.
(b) The slab is 1.5 cm thick. What is the displacement for red light incident at an angle of 45°? ($\lambda = 650$ nm).

P16.7. If you tune your radio to an FM station at 98.6 MHz, what is the wavelength? If the wave has to travel through water, what is its wavelength in water ($n = 1.33$)?

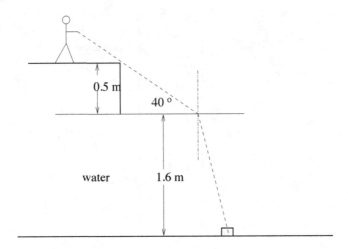

Fig. 16.12   Figure for Problem 16.12.

P16.8. You are speeding toward a traffic light that turns yellow. At what fraction of the speed of light must you travel for the light frequency to be shifted up by 20% so that it still looks green?

P16.9. Show that Eq. (16.5) is correct for small angles of incidence and refraction.

P16.10. A light ray propagates at 15° to the normal in a substance and emerges in the air at 40° with the normal.
(a) What is the index of refraction of the material?
(b) What is the critical angle for this material?

P16.11. (Challenge) The intensity of sunlight is 1390 W/m² at a distance of 1.5×10^{11} m from the sun. Find the total energy emitted by the sun per unit time.

P16.12. A friend loses her keys in the water of a lake. The keys fell from a dock 0.5 m above the water, which is 1.6 m deep near the edge of the dock. You search for the keys with a flashlight, as shown in Fig. 16.12, which just hits the edge of the dock, and is aimed at an angle of 40°, as shown in Fig. 16.12. The refracted light hits the keys in the water. Assume the index of water to be 1.33.

(a) What is the angle of refraction of the flashlight beam?

(b) How far below the water do the keys appear to be? (Note that the beam first travels in air).

(c) How far from the edge of the dock would you find the keys?

P16.13. Police use radar guns and the Doppler effect to catch speeding automobiles. The electromagnetic frequency of the gun is $f_s = 8.0 \times 10^9$ Hz. A car approaches the parked police car. The electromagnetic wave from the radar gun is reflected from the car and the frequency appears to the police at a frequency higher than f_s by 2100 Hz. What is the speed of the car?

Answers to odd-numbered questions

Q1. Black, unless looking directly to the sun, because there is no atmosphere.

Q3. Refraction of light. Yes, it would also be true at sunrise.

Q5. The light would be internally reflected; the angle of refraction would be larger than $90°$.

Q7. Blue and red.

Q9. It also decreases, since the electric field produces the magnetic field.

Q11. Red light travels faster, because the index of refraction is slightly smaller.

Q13. Yes, if there are two equal beams traveling in opposite directions.

Answers to odd-numbered problems

P1. 3×10^8 m/s $\times$ 365 days/y $\times$ 24 h/day $\times$ 60 min/h $\times$ 60 s/min $= 9.46 \times 10^{15}$ m.

P3. $6.38 \times 10^6 \times 2$ m/$(3 \times 10^8$ m/s$) = 4.25 \times 10^{-2}$ s.

P5. 1.92 m $\times 1.33 = 2.55$ m.

P7. $\lambda = 3 \times 10^8$ m/s/$(98.6 \times 10^6/\text{s}) = 3.04$ m. If the wave travels through water, $v = c/1.33$, and $\lambda = 3.04$ m$/1.33 = 2.28$ m.

P9. This follows from geometry and the fact that for small angles, $\sin\theta = \tan\theta$.

P11. $4\pi R^2 I = 4\pi(1.5 \times 10^{11} \text{ m})^2 \times 1390$ W/m$^2 = 3.93 \times 10^{26}$ W.

P13. $8 \times 10^7 \sqrt{\frac{c+v}{c-v}} = (8 \times 10^9 + 2100)$ Hz. From this it follows that $v/c = \frac{2100}{8 \times 10^9} \approx 2.6 \times 10^{-7}$, and $v = 78$ m/s.

Chapter 17

Geometrical Optics — Image Formation

In this chapter we shall study some applications of electromagnetic waves; we will examine reflection and transmission of light from mirrors and lenses.

17.1 Reflection from a Mirror

In *geometrical optics* light can be considered to travel in straight lines and rays are useful. These rays are like light beams from the source. If a plane silvered surface (mirror) is in the way, the reflected ray emerges at the same angle as the angle of incidence, as shown in Fig. 17.1. It is important to note that angles are measured from the normal to the surface, rather than from the surface. Thus, we have

$$\theta_r = \theta_i \,, \tag{17.1}$$

where θ_i is the incident angle and θ_r the reflected one. This equation is reputed to be the oldest recorded law of physics; it is attributed to Hero of Alexandria (10–70 AD).

It can be shown that this condition ($\theta_r = \theta_i$) corresponds to the shortest distance the light can travel from A to B via the mirror. Mirrors form images from an object. The image is where the reflected light *appears* to come from. You can use any two rays to locate the image and find that the image is as far behind the mirror as the object is in front, as shown in Fig. 17.1. The image is said to be *virtual* in this case, because the light does not actually go though the image; it only appears to do so. The image is right side up; it is not inverted. The image is the same size as the object. If the object is far away from the mirror, the image may look smaller, because the angle subtended at the eye is smaller than if it were near the mirror. If

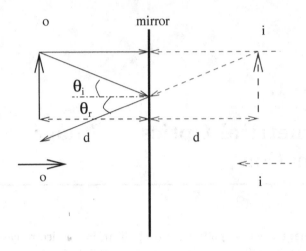

Fig. 17.1   Reflection from a mirror.

an object points towards the mirror, it is from left to right, but the image is from right to left, as shown in Fig. 17.1. If you look into a mirror, your right eye will appear on the left and your left eye on the right. Left and right are inverted.

What happens if you have several mirrors, such as two mirrors facing each other? In such cases, the image of one mirror serves as the object for the second one. For the situation just described, there will, in general, be an infinite number of images, further and further away from the mirrors. Another situation of interest is reflection from a corner mirror, with the two mirrors at 90° to each other. In that case there will generally be three images.

Parabolic concave mirrors (concave means that the silvered surface is the inside of a parabolic surface) reflect light from far away to a *focal point*, F, as shown in Fig. 17.2. Such mirrors are used as light collecting telescopes; the light is focused onto a film or other device for observing faint sources of light, such as that from a distant star. The reverse situation is used in a flashlight.

A spherical concave silvered surface also focuses light, but, unless the height of the object is small compared to the mirror's radius, the focus is not sharp. The focal length is the distance from the center of the mirror, along the axis to the focus is $\frac{1}{2}r$, where r is the radius of the spherical surface. By contrast, a spherical convex surface (like the outside surface

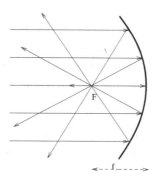

Fig. 17.2   Concave parabolic mirror.

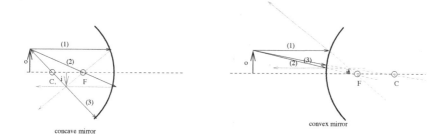

Fig. 17.3   An object in front of a concave and a convex spherical mirror.

of a sphere) defocuses the incident light. The focal point in that case is behind the mirror and is virtual. Objects placed in front of any mirror form images which can be located by means of rays. The image may be real, virtual, the same size, magnified or reduced in height. It depends on the location of the object relative to the focus and the shape of the mirror. Figure 17.3 shows an object placed beyond the center of concave and convex spherical mirrors, and three rays emanating from them. A ray, (1), parallel to the axis is reflected through the focus, if the object is not too tall. A ray, (2), through the focal point comes out parallel, and a ray, (3), through the center of the spherical surface is undeflected, since it is perpendicular to the surface. In all cases the angle of reflection is equal to the angle of incidence. An object placed beyond the center of a concave mirror gives rise to an inverted, real, and smaller image. An object similarly placed in front of a convex mirror leads to a virtual image, right-side-up and smaller.

17.2 Lenses

Images can not only be formed by reflection, but also by transmission through lenses. Eyeglasses are a prime example. As for non-plane mirrors there are both converging and diverging lenses. In general, a convex lens focuses light whereas a concave lens diverges it. For simplicity, we assume thin lenses, with the same curvature for both sides of the lens. For a thin lens, one can assume that the refractions all occur in the center of the lens. An example is shown in Fig. 17.4, where a converging lens focuses light rays parallel to the axis. E.g., light from far away to a focal point, F. The focal point is symmetrical about the lens since the lens is symmetrical between left and right. Its distance from the center of the lens is called the *focal distance, f*.

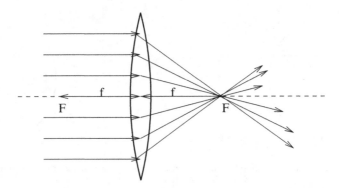

Fig. 17.4   A thin converging lens focuses parallel rays to the focus.

There are three rays that are useful for finding images, as shown in Fig. 17.5, but only two of them are needed to locate the image. For a converging (diverging are listed within parentheses) lens:

(1) a ray through the center of the lens is undeflected,

(2) a ray that comes in parallel to the axis of the lens passes through the focal point on the far side of the lens (on the object's side of the lens), and

(3) a ray through the focal point on the same (far)side of the lens as the object emerges parallel to the axis.

The image is real in Fig. 17.5 because the light actually goes through the image. Note that the image is on the other side of the lens, because

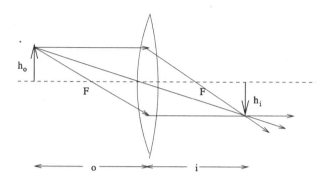

Fig. 17.5 A converging lens produces a real, magnified, image from an object placed further than the focus F.

reflection from the lens is neglected. If the object were placed closer than F, the image would be virtual. From geometry we can locate the image and find out its magnification. The formulas are (we will not derive them here:)

$$\frac{1}{o} + \frac{1}{i} = \frac{1}{f}, \tag{17.2}$$

$$\frac{h_i}{h_o} = -\frac{i}{o}, \tag{17.3}$$

where o, i, f are the object, image and focal distances from the lens and h indicates the height of the image or object. These equations are often called *lensmaker's equations*. They apply equally well to mirrors. The focal length of a plane mirror is infinite.

Signs are important in the lensmaker's equation. The focal distance f is positive for a converging lens and negative for a diverging one. If the image is the same side as the object, then i is negative; otherwise it is positive. A negative image distance implies a virtual one, since the image is on the same side of the lens as the object, and light does not actually go through it.

Illustrative Problem 17.1: As an example consider a converging lens with $f = 3$ cm and $o = 2$ cm. Find the image by rays and by the lensmaker's formula, as well as its height. Is it right-side up?

Solution: Application of the lensmaker formula gives $1/i = 1/f - 1/o = 1/3 - 1/2 = -1/6$. The image is thus at -6 cm, meaning that it is on

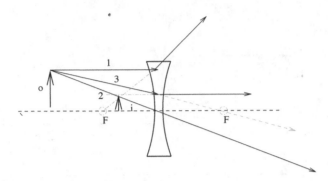

Fig. 17.6   Image formation from a concave lens (note that ray 3 heads towards the focus on the far side of the lens from the object).

the same side as the object and 6 cm away from the lens. The image is virtual. Its magnification is $h_i/h_o = -i/o = -(-6 \text{ cm})/2 \text{ cm} = +3$. Thus the virtual image is real side up (a negative sign would mean the image is upside down) and is three times as large as the object. It is worthwhile to draw the picture in order to check these results.

A diverging or concave lens makes virtual images that are right side up and generally smaller, as shown in Fig. 17.6. For a concave or diverging lens, the focal distance is negative.

Illustrative Problem 17.2: Let's do the same example as before, but for a diverging lens. Now $f = -3$ cm and $o = 2$ cm.

Solution: We find $1/i = 1/f - 1/o = -1/3 - 1/2 = -5/6$, or $i = -6/5 = -1.2$ cm. The image is virtual (on the same side as the object). The magnification of the image is $h_i/h_o = -i/o = -(-1.2)/2 = +0.6$. Thus, the image is smaller than the object and is virtual, but right side up. Again, draw the picture to check.

It is always a good idea to draw a picture. It serves to check that you have used the right signs in the lensmaker's equations and have not made a numerical error.

Combinations of converging and diverging lenses are used in microscopes and telescopes. The image from the first lens serves as the object for the second one. Figure 17.7 shows the basic layout for a microscope. It consists of two converging lenses; the one nearer to the object is called the *objective*

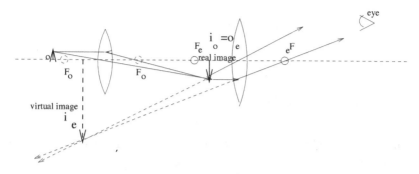

Fig. 17.7   Basic layout for a microscope.

lens and the one nearer the eye is the *eyepiece* or ocular. The objective has a relatively short focal length and the object to be studied is placed just beyond it. It forms a real, inverted, and enlarged image, which falls inside the focal distance of the eyepiece. The second image is then virtual, inverted and greatly magnified. The eyepiece itself works like a magnifying glass.

For a near sighted person, light from far away is focused ahead of the retina of the eye. This person needs an additional diverging lens to focus this light on the retina. For a far-sighted person, the problem is the opposite. Distant light is focused behind the retina and the individual needs a converging lens to focus the light onto the retina.

17.3   Light Polarization

Huygens believed that light was like sound, a longitudinal wave, and needed a medium to propagate. Newton believed that light behaved like particles. Neither one was correct, in that light can behave either as a wave or as particles, as we shall see. One of the features of a transverse wave is that it can be *polarized*. An example is a wave on a rope. If the rope is moved up and down, the polarization is in that direction. Such a wave could not pass through a horizontal slit, but only a vertical one; see Fig. 17.8.

By convention, the direction of polarization of light is that of the electric field. It can be in any direction perpendicular to the propagation of the wave. If no direction is preferred, the light is said to be unpolarized. Light coming from most sources consists of many waves with different directions of polarization and thus is unpolarized. If the electric field is in a definite

Fig. 17.8   A vertically oscillating rope and horizontal and vertical slits.

direction, the light is said to be *linearly polarized*. In some sense, therefore, polarized light is simpler than unpolarized light.

How can you produce linearly polarized light? Certain crystals split incoming light into two rays with perpendicular polarizations. Light can also be polarized by reflection. When light is reflected from a mirror or from, say water, the reflected beam of light is preferentially polarized in the plane parallel to the surface of the reflecting substance. The degree of polarization depends on the reflecting angle. At normal incidence, the reflected beam is unpolarized. It reaches its maximum (100%) polarization at *Brewster's angle*, see Fig. 17.9. Brewster's angle refers to the incident angle θ_i

$$\tan\theta_{Brewster} = n_2/1\,, \qquad\qquad (17.4)$$

where we have assumed that the incident light is traveling in air and is reflected from e.g., water with an index of refraction $= n_2$. In Fig. 17.9, θ_i is the incident angle, θ_r is the reflected angle and θ_R is the refracted angle in the substance of index n_2. Combining Brewster's law with Snell's law, we see that the reflected beam must be at 90° relative to the refracted one.

In 1929, Edwin Land made a plastic material containing aligned needle-like crystals, now called *Polaroids*. An electric field parallel to the long axis is absorbed, so that only light polarized perpendicular to this axis is transmitted. If an unpolarized beam of light is incident on this *polarizer*, the transmitted beam is linearly polarized perpendicular to the axis of the molecules. This is a simple and effective manner to achieve linearly polarized light. If you now place a second sheet of Polaroid, called *analyzer*, at right angles to the polarizer, then no light gets through at all. Neglecting absorption, the intensity of the light emanating from the first polarizer is only 50% of the incident one, since one direction of polarization is lost. If the angles between the polarizer and analyzer is α, then the intensity of the light coming out of the analyzer is

$$I_2 = I_1 \cos^2\alpha = \frac{1}{2}I_0 \cos^2\alpha\,. \qquad\qquad (17.5)$$

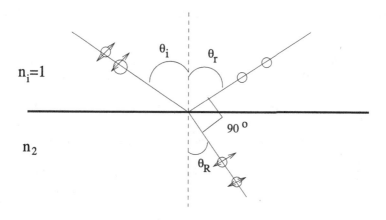

Fig. 17.9 Polarization of light by reflection and Brewster's law.

Here I_1 is the maximum intensity of light coming out of the polarizer and I_0 is the incident intensity. $I_1 = \frac{1}{2}I_0$ since the perpendicular polarization is absorbed.

What happens if you insert a third sheet of polarizer between the first and second ones? (See problems). Sunglasses use polaroid materials and are polarized perpendicular to the horizontal direction, since the reflection from puddles of water, lakes, and even roads tends to be polarized parallel to these surfaces, as "explained" by Brewster's law.

Illustrative Example 17.3: Light of intensity $= 3.2$ W/m^2 is incident on a series of polaroid sheets. The second sheet is aligned at $60°$ relative to the first sheet. What is the intensity out of the 2nd sheet?

Solution: $I_1 = 0.5 \times 3.2$ W/m$^2 = 1.6$ W/m^2. $I_2 = 1.6 \cos^2 60° = 0.4$ W/m^2.

Questions

Q17.1. Draw the wavefronts for light incident on a plane mirror from far away and the reflected light. Show these wavefronts close to the mirror.

Q17.2. Show how you would use a 45° prism to turn light around (by 180°).

Q17.3. Can a diverging lens produce a real image? If so, how; if not, why not?

Q17.4. Can a real image be projected onto a screen? A virtual image? Explain.

Q17.5. When will a spherical concave mirror give rise to a real image? Will it be smaller? Will it be inverted? Give reasons for your answers.

Q17.6. (a) Is the image of an object in a spherical concave mirror magnified or reduced in size? Explain.
(b) Is the answer to (a) dependent on the location of the object? If so, what is this dependence? If not, why not?
(c) When will a spherical concave mirror give rise to a virtual image?

Q17.7. Give two methods that can be used to focus the sun's rays into a small spot so that you can start a camp fire.

Q17.8. Why do goggles work under water, whereas the naked eye makes objects look blurred?

Q17.9. What is the focal length of a plane mirror?

Q17.10. The waves from the simple antenna, shown in Fig. 16.3 propagate to the right. Would you expect these electromagnetic waves to be polarized? If so, in what direction? If not, why not?

Problems

P17.1. What is the shortest plane mirror that can show your entire height? Give the reason and show graphically.

P17.2. For the case that light is incident on a lake at Brewster's angle, show that the angle between the reflected and refracted rays is $90°$.

P17.3. A spherical concave mirror of radius 11 cm has a 3.0 cm object placed at 1/2 the focal distance. Find its image graphically and with the lensmaker's formula. Is the image real or virtual, right side-up or inverted? larger or smaller? Find the image's magnification.

P17.4. Find the primary and secondary images for the object shown in Fig. 17.10, placed in front of a $60°$ hinged mirror.

P17.5. A thin diverging lens has a focal length 4.2 cm. An object 3.1 cm

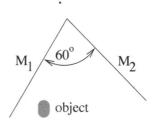

Fig. 17.10 Object in front of a hinged mirror.

high is placed 2.1 cm ahead of the lens. Find the image graphically and with the lensmaker's formula. Determine its properties: height, real or virtual, right side up or inverted.

P17.6. (a) An object 3 cm high is placed 10 cm in front of a lens of focal length = 6 cm. Find the image graphically and by the lensmaker's formula. Determine the height of the image. Is it real? Is it right side up?
(b) If the image serves as the object of a second lens with $f = +4$ cm, and placed 6 cm beyond the image serving as its object, locate the second image and its properties. Do both graphically and numerically.
(c) What is the overall magnification of the two lenses?

P17.7. (Challenge) Since water has an index $n = 1.33$, $\tan \theta_{Brewster} = 1.33$. What is the angle of refraction? (Hint: Look at Fig. 17.9).

P17.8. Show that the refracted ray comes out at 90° relative to the reflected one at Brewster's angle.

P17.9. A microscope has the two converging lenses separated by 6.0 cm. An object is placed 0.21 cm from the objective. The focal length of the objective is 0.20 cm and that of the eyepiece is 2.0 cm.
(a) Locate the image from the objective lens relative to the eyepiece.
(b) Locate the image from the eyepiece relative to the eyepiece.
(c) Determine the overall magnification.

P17.10. Consider three sheets of polaroid with their axes at 0°, 45° and 90°, respectively.
(a) Will some light come out of the third sheet? Explain.
(b) If your answer to (a) is yes, what is the intensity of the light beam emanating from the third sheet if the incident intensity is I_0? Assume no losses in light transmission through the polaroid sheets.

Answers to odd-numbered questions

Q3. No, because the rays diverge.

Q5. When the object is further away from the mirror than the focal length. The image is inverted and smaller. For reason, see the lensmaker's formula.

Q7. Use of a parabolic concave mirror or a convex lens.

Q9. Infinite.

Answers to odd-numbered problems

P1. Half your height so that light from your feet, reflected in the mirror, still hits your eyes.

P3. The focal distance is 5.5 cm and the object is placed at 2.75 cm. $1/i = 1/5.5 - 1/2.75 = -1/5.5$; $i = -5.5$ cm. magnification $= h_i/h_o = -i/o = 5.5/2.75 = 2$. The image is right side up and magnified, but virtual (it is behind the mirror).

P5. $i = -1.4$ cm; $h_i = 2.1$ cm.

P7. 37°.

P9. (a) The image from the objective is located at 4.1 cm from the objective and thus at 1.9 cm from the eyepiece.
(b) The image from the eyepiece is located at -38 cm from the eyepiece.
(c) The magnification of the objective is $-i/o = -4.1/0.21 = -19.5$. The magnification of the eyepiece is $-i$ (eyepiece)$/o$ (eyepiece) $= -38/1.9 = -20$. The overall magnification is therefore $20(19.5) = 390$.

Chapter 18

Physical Optics

In the previous chapter we studied geometrical optics. In this chapter we shall study physical optics phenomena, which differentiate waves from particles: interference and diffraction. *Diffraction* is the bending of light (or other waves) around opaque objects and corners. Sound waves do that, since we can hear voices in a room from a corridor.

18.1 Interference Phenomena

Perhaps the most definitive proof that light is a wave came from an experiment carried out by Thomas Young (1773–1829). Young was born into a Quaker family and learned 12 languages. He studied both medicine and physics and contributed to both subjects. In physics, he is best known for his double slit experiment (Fig. 18.1). Light from a source is incident on a plane that is solid except for two slits, placed symmetrically about the source, as shown in Fig. 18.1. The light is then projected onto a screen some distance from the slits. The distance between the slit openings, d, is of the order of 10–100 times the wavelength(s) of the light, and the slits are also of that order of magnitude in width.

The center of the screen (opposite the source) is found to be bright; the light has clearly bent towards this center. But there are other bright lines on both sides of the center and at some distance thereof. These bright fringes can be explained on the basis of constructive interference from the light passing through the two slits, whereas the dark lines are due to destructive interference. We can use Huygen's principle to understand these phenomena. The two slits are sources of new wavelets expanding to the right, and interfere. At the center of the screen the two waves arrive in phase, since

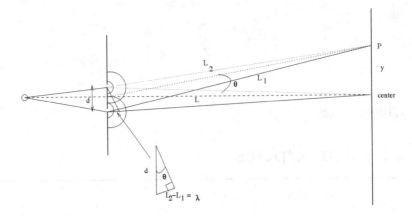

Fig. 18.1 The Young double slit experiment. We show constructive interference for $|L_2 - L_1| = \lambda$.

they travel the same distance. But at some distance, y from the center of the screen, at P, the difference in path lengths from the two slits is $\frac{1}{2}\lambda$ if the light has a definite wavelength. The two waves arrive out of phase at P and interfere destructively. Thus, P will be dark. Somewhat higher up (or down) the difference in distance will be an entire wavelength and Q will be bright. This continues at regular intervals and there will be dark "lines" whenever the difference in path lengths is $\frac{1}{2}\lambda$, $\frac{3}{2}\lambda$, $\frac{5}{2}\lambda \dots$, $\frac{(2n+1)}{2}\lambda$ and bright lines when the difference of path lengths is $n\lambda$, where n is an integer. We can determine the distance y, where these lines occur on the screen. From the inset of Fig. 18.1, we see that the difference in path length is

$$\Delta L = |L_2 - L_1| = d\sin\theta. \tag{18.1}$$

We also see that

$$\tan\theta = y/L, \tag{18.2}$$

where L is the distance from the plane of the slits to the screen. For small angles (please note the assumption) $\sin\theta \approx \tan\theta$, so that in this case we can find y for bright lines

$$\Delta L = |L_2 - L_1| = N\lambda = d\sin\theta \approx d\tan\theta = d\frac{y}{L}, \tag{18.3}$$

or

$$y = \frac{L}{d}N\lambda. \tag{18.4}$$

Here N is any integer. We see that the difference in path length, $n\lambda$ has been magnified by $\frac{L}{d}$. The center is always bright, and the dark fringes, as they are called, will be between the bright lines when

$$y = \frac{L}{d}\left(n + \frac{1}{2}\right)\lambda. \tag{18.5}$$

These equations hold only when the angles involved are small. If the slit separation is much larger than the wavelength, say 1000 times as large or larger, the separation of bright fringes will be too small to observe. You will not see interference in this case. On the other hand if the slit separation is (of the order of) one wavelength, you may only be able to see the central bright line. The first bright fringe on the side of the central one would occur at $\sin\theta = \lambda/d \approx 1$. Young's double slit experiment is a clear example of the superposition of waves and could not occur for particles.

Illustrative Example 18.1: (a) Find the vertical distance of the 2nd bright fringe on one side of the central maximum if $\lambda = 5200$ Å; the distance from the slits to the screen is 2.2 m and the separation of the slits is 0.22 mm. (b) Find the angle made by this 2nd maximum with the horizontal at the center of the slits.

Solution: (a) $y = N\lambda(L/d) = 2 \times (5200 \times 10^{-10}) \times 2.2/(2.2 \times 10^{-4}) = 1.04 \times 10^{-2}$ m $= 1.04$ cm.
(b) $\tan\theta = 1.04 \times 10^{-2}$ m/2.2 m $= 4.7 \times 10^{-3}$; $\approx \theta = 4.7 \times 10^{-3}$ rad.

18.2 Diffraction Gratings

It is also possible to use more than two slits. You can overcome some of the disadvantages of 2 slits in separating colors because the colors may not be separated sufficiently, and the maxima may be too wide. If you have multiple slits with equal separations, you can repeat the analysis for two slits. You can also have wire gratings that are equally spaced; indeed, the first *diffraction gratings* were constructed by Joseph Fraunhofer in 1820. He had as many as 19 wires/mm. In the 1880's, Henry A. Rowland succeeded in making gratings with as many as 1000 lines/mm. Nowadays, it is possible to make 4000 lines/mm; the spacings between two lines are then 250 nm. If the extra distance traversed by light from one "slit" to the next one is λ, then the following slits will have longer path lengths by $n\lambda$, where n is an integer, and a maximum intensity will occur on the screen. If the separation of neighbors differs slightly from one wavelength, the intensity drops sharply because the next slit is further out of phase and so forth. Thus, the screen interference patterns have sharper maxima than for two slits, which makes it easier to separate colors. Indeed, diffraction gratings are used as *spectrometers*, to measure wavelengths and separate colors.

You also get secondary maxima due to the constructive interference from slits more than one removed; for example, if the difference in lengths from neighboring slits is $(1/3)$ λ.

18.3   Diffraction from a Single Slit or Obstacle

Since light bends around obstacles that are not too large compared to the wavelength of the light, the central point on a screen beyond the tip of a needle, for instance, will be bright, since the light from both sides will traverse the same lengths and arrive in phase. This bright spot demonstrates the bending, or *diffraction* of light.

Diffraction also occurs for a single slit. Huygen's principle allows us to understand the bending of light around obstacles, as well as from slits.

To calculate the position of the first minimum, we divide the slit into two halves. If the path difference between the top of the slit and the upper part of the lower half of the slit to the screen is $\lambda/2$, there will be maximum destructive interference between the upper and lower parts of the slit. Each point in the upper half has a counterpart in the lower half for which the difference in path lengths will be $\lambda/2$. This is illustrated in Figs. 18.2 and 18.3, where the first minimum occurs at ± 0.2. For the first maximum on

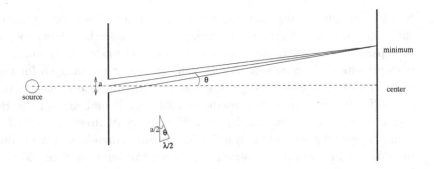

Fig. 18.2   Diffraction from a single slit. The conditions are shown for the first minimum on the side of the center.

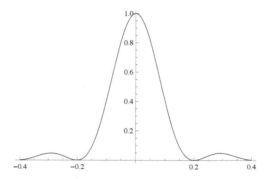

Fig. 18.3   Intensity for diffraction from a single slit.

either side of the middle we need a difference in path lengths of λ between each point of the upper and lower halves of the slit. Thus, the conditions are

$$\sin\theta = \frac{\lambda/2}{a/2} = \frac{\lambda}{a} \cdots \text{ for first } minimum, \qquad (18.6)$$

$$\sin\theta = \frac{\lambda}{a/2} = \frac{2\lambda}{a} \cdots \text{ for first } maximum, \qquad (18.7)$$

where a is the slit opening. The angles can be translated to screen distances in the same way as for the double slit. When you have two slits, you also have two single slits. Diffraction occurs for the single slits and interference from the two slits. Both are interference effects. The single slits will necessarily have a wider diffraction pattern than the interference from the two slits, since the separation of the two slits must be larger than the size of each one and the interference pattern depends inversely on these distances. The single slit diffraction pattern will thus modulate the interference pattern. The intensity is the square of the amplitude and provides an "envelope" for the interference pattern, as shown in Fig. 18.4.

Illustrative Example 18.2: If in Illustrative Example 18.1, the slits are each 0.05 mm in diameter, find the first minimum of the diffraction pattern.

Solution: $\sin\theta = \lambda/a = 5.2 \times 10^{-7} \text{ m}/(0.05 \times 10^{-3} \text{ m}) = 1.04 \times 10{-2} \approx \theta$; $y = L\tan\theta \approx 2.2(1.04 \times 10^{-2} \text{ m}) \approx 2.3 \text{ cm}$.

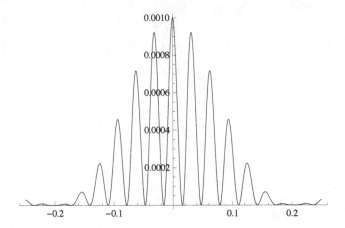

Fig. 18.4 Intensity for a two slit interference pattern modulated by the diffraction of each slit. The envelope of the two slit interference is that of Fig. 18.3. The interference pattern alone is not shown.

It is usually assumed that the smaller you make a pinhole the sharper will be its image on a screen or photographic plate. This is only true to a degree. As the pinhole gets to be smaller and smaller, diffraction begins to play a role and the first minimum of the diffraction pattern defines the sharpness of the picture. As a approaches λ, the first minimum shifts to larger and larger angles and the image gets less and less sharp.

18.4   Some Other Interference Phenomena

Other observable interference phenomena are those from reflection from a thin film of oil, water, or soap, from spherical lenses placed on a flat one, and from two glass plates separated by a small distance. In all cases the distances involved are fractions of millimeter and the films are looked at from straight above them, so that the angle of incidence is 0°.

Consider reflections from a thin oil film. Light is incident at 0° and is reflected from the top surface as well as from the bottom one. If the viewing angle is close to the normal, then the extra distance for the light reflected from the bottom is $2t$, if t is the thickness of the slick. If $2t$ is λ, there will be destructive interference and you will not see the film! The reason is given below. You must take care to remember that λ is the wavelength in oil and not in air!

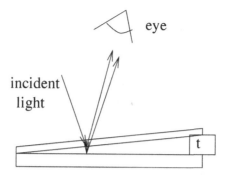

Fig. 18.5   Interference from light reflected from two glass plates separated by distance *t*. The interference is between light reflected from the bottom of the top plate and the top of the bottom plate. The distance *t* should be small compared to the thicknesses of the two glass plates. Light can be considered to be perpendicular to both plates.

Reflection from a soap film on water or from two thin glass plates separated by a tiny distance has an extra degree of difficulty. When the light is incident from a less dense medium to a denser one, e.g., air to glass, there is a phase change of 180°, equivalent to $\frac{1}{2}\lambda$. This is akin to a vibrating rope at a nail that fixes it to e.g., a wall. Thus, if the separation of the two relatively thick glass plates is $\frac{1}{4}\lambda$, the additional *optical* distance by the light reflected from the two glass surfaces is $\frac{1}{2}\lambda + \frac{1}{2}\lambda = \lambda$ and constructive interference occurs. Once again, the wavelength is that in the soap film ($n \approx 1.3$) and not in air; the angle of incidence can be considered to be 0°. See Fig. 18.5 for an illustration.

When you place a curved lens on a flat glass plate, the distance between the two glass surfaces increases with the angle relative to the axis of the curved surface. You can then see a circular diffraction pattern. The rings are called "Newton's rings". The center of the pattern will be dark. Can you tell why this is so? (See Questions).

Illustrative Example — Challenge: Newton's rings are formed between a hemispherical piece of glass of radius 4.2 cm and a flat plate of the same index, $n = 1.5$. What is the radius of the first bright fringe for red light of $\lambda = 680$ nm?

Solution: (See Fig. 18.6). The distance between the hemispherical glass and the plate is $h = R(1 - \cos a)$. The distance needs to be $\lambda/4$, so that $2h$ (the extra distance gone by one of the interfering beams) is $\lambda/2$; that extra length

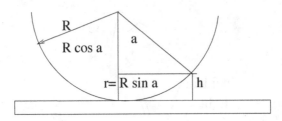

Fig. 18.6   Newton's rings.

added to the $\lambda/2$ phase change due to reflection of the beam from the flat plate of glass gives a path difference of λ. Thus, $h = 680$ nm/4 = 170 nm. For small angles $\cos a = 1 - a^2/2$; since $\cos a = 1 - 4.05 \times 10^{-6}$, we find $a = 1.4 \times 10^{-3}$ radian. The radius $r = R \sin a \approx Ra = 4.2 \times 1.4 \times 10^{-3}$ cm = 5.9×10^{-3} cm.

   In order to show the wave nature of X-rays we need apertures or obstacles of the order of one nm. Such openings or obstacles are difficult to come by, but Max von Laue realized in 1912 that atoms in a crystal lattice form a 3-dimensional diffraction grating. The separation of planes of atoms in a crystal is typically of the order of 0.5 nm. As shown in Fig. 18.6, the diffracted waves from adjacent planes must differ in path length by a whole number of wavelengths to get constructive interference. If the wave is observed by reflection, the angle ϑ with the horizontal is equal to the similar reflected angle, as shown in Fig. 18.7. If the spacing between adjacent planes is d, then constructive interference requires the path length difference to be $m\lambda$.

$$2d \sin \theta = m\lambda, \tag{18.8}$$

with m an integer, so that the path length of the two rays shown differ by an integer times λ. This condition is called *Bragg's law*. X-ray diffraction on crystals can be used to study crystal structure.

   Both compact disc (CD) and digital video disc (DVD) players work on the principle of interference. They hold information on a spiral track that is detected by a laser beam reflected from the bottom of the disc. The disc has raised areas, called pits, separated by flat ones, called land, and covered by a plastic coating. The laser beam reflects light from these areas. The pit thickness is chosen as half of a wavelength in the material, so that there is destructive interference at the edge (to reduce background noise). The reflection of the laser light gives a set of 0's and 1's.

Fig. 18.7 Bragg reflection from a crystal. The insert shows the condition for a maximum.

Questions

Q1. What is the shape of the bright bands on the screen for two or more rectangular slits? What about circular holes?

Q2. What is the shape of the bright fringes on the screen for the diffraction of light from a pinhole?

Q3. What happens to the separation of the first maximum (from red laser light) from the center of the projection of a single slit if the width is doubled?

Q4. The interference pattern observed from the reflection off a soap film is multicolored. Explain.

Q5. Why don't we observe interference between the front and back of eyeglasses?

Q6. Show that the diffraction pattern from the individual slits of a Young apparatus is wider than than the interference pattern.

Q7. Draw the interference pattern from two long slits that are separated by $\frac{3}{4}\lambda$.

Q8. Repeat question 7 for a separation of $\frac{1}{2}\lambda$.

Q9. Thin film interference occurs when one glass plate is placed on top of another one. What two waves are interfering in this case? Is there a phase shift associated with the reflection? If so, where?

Q10. Show graphically what happens as you narrow the size of an aperture when the opening gets close to one wavelength of the light. Draw the intensity of the light for several apertures close to λ.

Q11. If the separation between 2 slits is halved, what happens to the interference pattern on a screen? Explain.

Q12. List three properties that are particular to waves.

Q13. Newton rings have a dark, rather than a bright center. Explain.

Problems

P1. Two slits in a solid surface are 0.20 mm apart. They are illuminated by white light.
(a) Which color will have its first bright fringe closest to the central (white) one?
(b) What is the separation of red ($\lambda = 680$ nm) and blue light ($\lambda = 460$ nm) at this first bright fringe on a screen 2.0 m away?
(c) What is the separation on the screen for the blue light at the second maximum, beyond the central one, and red light at the first maximum?

P2. Two glass strips of length 15 cm are separated by a piece of paper at one end. The paper is 0.62 mm thick. How many bright interference fringes would you observe with mercury light of $\lambda = 546$ nm? Assume normal incidence. Do you need to worry about a phase change at one surface? If so, which one?

P3. Two slits are 3 wavelengths apart. What pattern will appear on a screen placed 2 m away. Find the locations of maxima and minima on the screen. How many of each are there?

P4. A diffraction grating has 2000 lines/cm. It is illuminated by a mercury lamp of wavelength = 546 nm. What is the angular separation of bright lines?

P5. A single pinhole has a diameter of 0.22 mm. Light of wavelength = 600.0 nm is incident on the pinhole. What will be the width of the central maximum? That is, where does the first minimum occur?

P6. Reflection occurs from an oil film 0.020 mm thick on water. The index of refraction of the oil is 1.40 and that of water is 1.33. How many minima will there be at an incident wavelength of 500.0 nm?

P7. Green ($\lambda = 520$ nm) and red ($\lambda = 680$ nm) light are incident on a soap film ($n = 1.33$) with air on either side. What is the minimum thickness for the film to appear green?

P8. Two sound speakers are placed 1.0 m apart. You are located at an angle of 20° relative to the axis (midpoint between the speakers). What sound frequency or frequencies will you not be able to hear? Take the speed of sound as 330 m/s.

P9. What is the smallest diameter of a single pin hole to be able to focus light from a mercury lamp ($\lambda = 546$ nm)?

P10. Instead of a single slit, we have diffraction around a small sphere of radius 0.020 mm. What is the angle of deflection for red light ($\lambda = 680$ nm) at the first maximum?

Answers to odd-numbered questions

Q1. Rectangular interference patterns; circular interference patterns.

Q3. The separation is halved.

Q5. The eyeglasses are too thick.

Q7. The first minimum occurs at $\sin\theta = 4/6$ and the first maximum at $\sin\theta = 4/3$. Thus only the central maximum will be visible.

Q9. The interference occurs between the bottom of the first plate and the top of the second one. There is a phase shift of $1/2$ wavelength at the top of the second plate.

Q11. If the separation gets halved the images get twice as far apart, since $y = \frac{L}{d}\lambda$.

Answers to odd-numbered problems

P1. (a) Blue light, since it has the shortest wavelength.
(b) $y = (L/d)\lambda = (2 \text{ m}/0.2 \text{ mm})\, 680 \text{ nm} = 680 \times 10^4 \text{ nm} = 6.8 \text{ mm}$.
(c) $y_{blue} = 2(L/d)\, 460 \text{ nm}$, $y_{red} = (L/d)\, 680 \text{ nm}$. Separation $= (L/d)(920-680 \text{ nm}) = 2/(0.2 \times 10^{-3})\, 240 \text{ nm} = 2.4 \text{ mm}$.

P3. $\sin\theta = N\lambda/3$. The first bright fringe away from the middle would be at $\sin\theta = 1/3$, the 2nd one at $\sin\theta = 2/3$ and the third one at 90°. This one will be hard to see, but you cannot see more than 3 bright lines on either side of the central maximum. The dark ones will be at $\sin\theta = 1/6$, $1/2$, $5/6$. On the screen, these lines or dark areas appear at $y = L\tan\theta$, but $\sin\theta \neq \tan\theta$. Including the bright line at 90°, there are 7 bright lines and 6 dark spots.

P5. The first minimum occurs at $\sin\theta = \lambda/a = 2.7 \times 10^{-3}$.

P7. There is no phase change from air to soap, but there is one of 180° from the soap to air. You need $2t = \lambda/2$, $t = \lambda/4$ for 520 nm, or $t = 130$ nm. To see red light you would need a larger thickness.

P9. $a \leq \lambda = 546$ nm.

PART 3
Modern Physics

Chapter 19

The Beginning of Modern Physics

19.1 Blackbody Radiation

We have studied electromagnetic waves and some applications thereof in the previous two chapters. There are many other applications. One form of electromagnetic radiation is emitted when a charged particle is accelerated or decelerated. This form of radiation has acquired the German name *bremsstrahlung*, meaning slowing down radiation. It is used for identifying charged particles (the radiation depends on the mass of the particle) and also in the study of solid bodies. The radiation is produced by accelerating electrons in a circular accelerator, where there is a centripetal acceleration, which causes the electrons to radiate (lose) part of their energy. These accelerators have multiplied in the past decades and now exist all over the world.

Another form of electromagnetic radiation is that emitted by a hot body. It is often called *thermal radiation*. The sun is a prime example. The frequency distribution of such radiation tends to be in the infrared region if the temperature is not too high, and depends on both the temperature and the material of the emitter. An example of the wavelength distribution for two temperatures is shown in Fig. 19.1. The power or intensity dependence shows a maximum. As the temperature increases, the maximum shifts to higher frequencies, i.e., shorter wavelengths. The German physicist Wilhelm Wien found that the wavelength of the maximum is inversely proportional to the temperatures, or $\lambda_{\max}T = \text{constant}$,

$$\lambda_{\max}T = 2898 \ \mu\text{m-K} = 2.898 \times 10^{-3} \ \text{m-K} \qquad (19.1)$$

Wien was awarded the Nobel prize in 1911 for his research in thermal radiation.

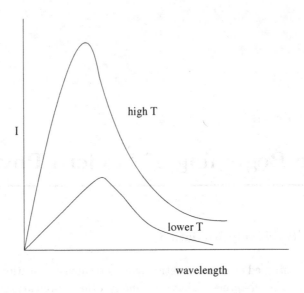

Fig. 19.1   Spectral distribution of emissivity from a hot body.

The sun's surface temperature is about 5800 K, so that the maximum in the power radiated occurs at a wavelength of about 500 nm, in the yellow region.

There is a method which allows one to obtain thermal radiation that is independent of the material of the emitter. You take a box of a material, e.g., a metal, which reflects the radiation kept inside it. The box is heated and kept at a constant temperature. The radiation emitted by the walls is reflected myriads of times inside the cavity. The cavity is made to have small hole, so that some of the radiation can escape if it has the right direction. Such a cavity radiator is said to be a *black body.* Any material can be used to contain the radiation. The distribution of wavelengths is shown in Fig. 19.1. An approximate example is the hot fire in a fireplace. When you look at the glow between logs, you tend not to see any details, i.e., whether it is wood, iron, or brick.

The distribution of wavelengths at a given T is universal. If the curve is universal, there must be an equation which describes it. Max Planck, a German physicist, and in some sense, the father of modern quantum physics, derived a formula for it.

$$I(\lambda, T) = \frac{2\pi c^2 h}{\lambda^5} \frac{1}{e^{hc/(\lambda k_B T)} - 1}. \tag{19.2}$$

Here I is the intensity of the radiation, k_B is the Boltzmann constant ($k_B = 1.38 \times 10^{-23}$ J/K) and h is a new universal constant, now called Planck's constant; its value is $h = 6.626 \times 10^{-34}$ J-s. This is the first introduction of this new constant, which plays an important role in quantum mechanics. Because it is small, it matters primarily at very small distances. Planck was forced to make the daring hypothesis that electromagnetic radiation is not emitted continuously, but only in discrete quanta, each with an energy

$$E = nhf = nhc/\lambda, \tag{19.3}$$

where n is an integer and f the frequency of the radiation. Planck called his hypothesis "an act of desperation"; it was the only way he could fit the spectrum over the entire range of frequencies. Planck assumed that the atoms in the cavity wall behaved like tiny oscillators emitting radiation at characteristic frequencies. The fact that energy was quantized was indeed a radical departure from classical physics.

The Planck formula for the distribution of power in black-body radiation applies in many situations, including the early years of the universe. The universe was, we believe, created by a very hot "big bang" and has cooled ever since. The universe became transparent to light when the light could propagate far without being absorbed. This transition happened about 10^5 years after the big bang, when the temperature of the universe had fallen to about 3000 K. The universe has continued to cool and expand, and today's world emits black body radiation at a characteristic temperature of only 2.73 K. This allows us to infer the age of the universe as 1.37×10^{10} years. It was discovered by A. Penzias and W. Wilson when they found a pervasive radiation from the sky that they could not, initially, understand. It is now called *cosmic microwave background radiation*. The peak of the spectrum lies in the microwave region of frequencies. Further studies of the radiation have revealed details about the very early development of the universe. For example, astrophysicists have been able to infer the existence of a very short duration (about 10^{-32} s) period during which the universe expanded exponentially, its size doubling every 10^{-34} s.

19.2 The Discovery of the Electron

Although Planck's formula for the radiation emitted by a black body may have signaled the birth of quantum mechanics, modern physics can be said to have begun with the discovery of the electron.

Fig. 19.2   A gas discharge tube.

The discovery was preceded by the invention of a more powerful vacuum pump, which allowed studies of gaseous emission from partially evacuated tubes. The tubes contained two plates connected to a battery, as shown in Fig. 19.2. The plate connected to the negative terminal of the battery is called a *cathode* and that connected to the positive one is called an *anode*. A glow could be seen in the tubes between the cathode and anode; the color was characteristic of the gas used in the tube. As the pressure is reduced in the tube, the glow is replaced by a dark column that leaves a (green) glow on the far end of the tube glass, opposite the cathode, near the anode. By repeated experiments, Crookes and others convinced themselves that the deposit came from the cathode. The color did not change with different gases or different cathode materials. The rays were called *cathode rays* and traveled in straight lines from the cathode. Further, it was found that magnetic fields affected the rays, so that it was concluded that they must be charged particles. In 1897, J. J. Thomson showed that both electric and magnetic fields affected the beams and that he could produce conditions, as in Fig. 19.3, such that no deflection of the beam took place. In this case we have

$$F = qE - qvB = 0 \tag{19.4}$$

The "−" sign in the second part of the equation is because the magnetic force was chosen to oppose the electric one, so that the net force vanishes. Thomson realized that the case of no deflection allowed him to obtain the velocity of the charged particles,

$$v = E/B \tag{19.5}$$

By measuring the trajectory with either force alone, Thomson could obtain the ratio q/m, since the acceleration is $a = F/m$, e.g. for a magnetic field qvB/m. In this manner, he was able to deduce that the ratio of q/m was about 1800 times larger than for hydrogen ions. Either the charge was

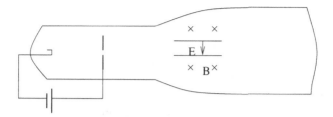

Fig. 19.3 Thomson's experiment for the discovery of the electron.

much larger or the mass much smaller. From earlier experiments by others, he showed that the charge was about that of a hydrogen ion. He thus surmised that the mass was much smaller, and so is hailed as the discoverer of the electron. Electrons are emitted by many atoms and thus must be a part of many substances. Electrons are also responsible for the chemical properties of substances. They were the first "fundamental" particle of modern physics to have been found.

19.3 The Electric Charge of the Electron

The American physicist Robert A. Millikan figured out a clever way to measure the charge of the electron directly. He squirted oil drops into a closed chamber by means of an atomizer. He studied the motion of these drops. Since there is only gravity and a viscous drag (frictional) force, the drops reach a terminal speed when the two forces are equal and opposite. Millikan then applied an electric field to the drops, which had been charged by friction when they left the atomizer. If the electric field is downward, the force on a negatively charged particle is upward and opposes gravity. If the forces could be made equal, the drops would not move. That is not easy, but the drops will acquire new terminal speeds due to the extra force, and Millikan found by a series of experiments that all the drops had charges of $q = ne$, where n is an integer and e is the smallest charge ($n = 1$). For e he obtained

$$e = -1.60 \times 10^{-19} C.$$

(19.6)

This not only fixes the charge of the electron as being very small, but also says that all charges are integer multiples of e. Millikan received the Nobel prize in Physics in 1923 for this work.

19.4   The Photoelectric Effect

In the early 1900's it was well known that heating a cathode could produce electrons. The electrons could be collected by placing an anode at a positive voltage relative to the cathode in the tube. It was found some time later that electrons could also be ejected from some metal cathodes by shining light on it. This phenomenon was studied quantitatively in the early 1900's. The results were difficult, indeed, to understand at that time. A typical apparatus is shown in Fig. 19.4. The current produced is measured by an ammeter.

Here are some of the results that were obtained:

(1) For certain colored lights, no current was obtained. The same was true for certain metal cathodes.
(2) The current measured, if any, was proportional to the intensity of the light source.
(3) Only when the frequency of the light was high enough was a current obtained.
(4) The electrons were emitted almost instantly, no matter how small the intensity of the light source.
(5) When the polarity of the external battery was reversed, so as to slow down negatively charged particles, it was found that no current flowed for a sufficiently high V, called *stopping voltage.*
(6) The stopping voltage was independent of the intensity of the light, but varied linearly with frequency of the light, as shown in Fig. 19.5. The lowest frequency for obtaining a current, the *threshold* frequency was f_0, with $hf_0 = W$.

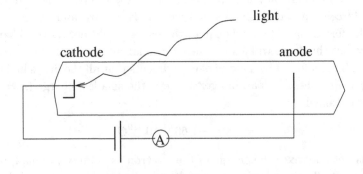

Fig. 19.4   The apparatus for studying the photoelectric effect.

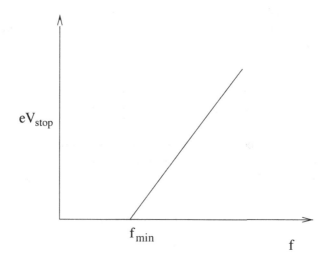

Fig. 19.5   The stopping voltage dependence on frequency of the radiation.

Einstein explained the photoelectric effect in 1905 (called the miraculous year, because he made 3 important discoveries that year). The crucial assumption was that light comes in wave packets of energy given by

$$E = hf, \qquad (19.7)$$

where h is Planck's constant. The energy of each *photon* is independent of the intensity of the light, but only depends on the frequency. This explains why a minimum frequency is necessary for the photoelectric effect. Electrons are emitted almost instantly from atoms, independent of the intensity of the light. It takes a certain amount of energy to wrest an electron from its normal place. In a metal, the minimum energy required is called the *work function, W*. If $hf_{\min} = W$, then the electrons are emitted with no kinetic energy and cannot reach the anode, unless it is at a positive voltage relative to the cathode. If $hf \geq hf_{\min} = W$, the excess energy goes into kinetic energy of the electrons,

$$K_{\max} = hf - W. \qquad (19.8)$$

We have used $K_{\max}$ because not all the energy of the photons may be given to a single electron and an electron may lose some energy before leaving the metal. The stopping voltage can also be understood. The current is linearly dependent on the intensity of the light beam, which is directly proportional to the number of photons. The potential energy $|e|V_{stop}$ is just equal to the maximum kinetic energy of the electrons, since they can only go to the top of a hill with a potential energy difference $\leq$ than their maximum kinetic energy. Thus

$$eV_{stop} = K_{\max}. \tag{19.9}$$

From Eqs. (19.8) and (19.9) we see that the slope of the curve in Fig. 19.5 is just equal to h and the intersection with the abscissa (x-axis) is equal to W.

It should be noted that in atomic physics and subatomic physics, energies are often measured, and given, in electron-volts, or eV. One electron volt is the energy gained or lost by an electron by falling through a potential difference of 1 V,

$$1 \text{ eV} = e \times 1 \text{ V} = 1.6 \times 10^{-19} \text{ J}. \tag{19.10}$$

In these units the Planck constant is

$$h = 4.135 \times 10^{-15} \text{ eV-s}. \tag{19.11}$$

Illustrative Problem 19.1: If the work function $W = 2.4$ eV, what is the minimum frequency of light to get electrons out of a cathode by shining light on it? If the frequency is twice the minimum, what is $K_{\max}$ and the stopping voltage?

Solution: If we have doubled the frequency, the $K_{\max} = 2.4$ eV and the stopping voltage is 2.4 V.

Robert Millikan showed the correctness of Einstein's explanation in 1915 and from the slope (see Fig. 19.5) was able to accurately measure Planck's constant h. This was, as he said (in 1948) "in spite of its unreasonableness, since it seemed to violate everything we knew about the interference of light." Even Einstein himself had qualms and in 1911 he still insisted on the provisional nature of the quantum concept because "it does not seem reconcilable with the experimentally verified consequences of the wave theory ..." But, it is the explanation of the photoelectric effect which earned Einstein the Nobel prize.

We have come to a crossroad. We have found definitive behavior of light and other electromagnetic radiation that shows that light behaves as waves. We have now seen two experiments, black-body radiation and the photoelectric effect which cannot be understood unless electromagnetic radiation is made up of packets of particles, called *quanta*. Which is correct?

It turns out that both are right and that light can be said to be schizophrenic or multi-faced. If you do experiments to demonstrate that light is a wave, it will indeed show behavior like one. If you do experiments to show that light behaves as quanta, you will find that it behaves that way. It all depends on the experiments you do! Quantum mechanics teaches us that both pictures are correct.

19.5 Compton Scattering

Arthur Holly Compton, another American physicist, set out in 1923 to measure the scattering of photons from electrons. He used X-rays, which are more energetic than visible light. If light indeed behaves as photons, it ought to collide with electrons just as two billiard balls. The photons could rebound in various directions, but, for an elastic collision, both energy and momentum must be conserved. If the initial energy of the photons is $hf = hc/\lambda$, then their momentum would be $E/c = hf/c = h/\lambda$. We will see this relationship again in the chapter on relativity. Thus, for an electron initially at rest, energy conservation requires

$$K'_e + hc/\lambda' = hc/\lambda, \tag{19.12}$$

where the primes refer to the final values of the variables and K is the kinetic energy. The electrons are assumed to be initially at rest. For photons that come out at 180° relative to the initial ones, momentum conservation requires

$$-h/\lambda' + p'_e = h/\lambda, \tag{19.13}$$

where p'_e is the momentum of the recoiling electron ($p'_e = \sqrt{2m_e K'_e}$). We can solve for λ' from energy and momentum conservation and find

$$\lambda' = \lambda + 2h/(m_e c) = \lambda + 2 \times 2.4 \times 10^{-12} \text{ m}. \tag{19.14}$$

This relationship holds only for backward scattering of the photons. The length $h/m_e c = 2.4 \times 10^{-12}$ m is now known as the *Compton wavelength of the electron.*

So, the Compton effect is another experiment that can be understood for electromagnetic radiation composed of particles, photons. We have now seen three experiments that require us to consider light as streams of particles:

(1) the blackbody radiation,

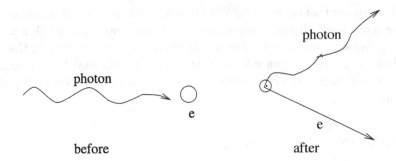

Fig. 19.6   Compton scattering from an electron.

(2) the photoelectric effect, and
(3) the Compton effect.

We will see in Chapter 21 how to reconcile the particle and wave pictures of electromagnetic radiation.

Illustrative Question: For elastic Compton scattering, is the frequency of the photon after the collision equal to that prior to the collision? Explain.

Answer: No, the wavelength changes and thus the frequency also changes, depending on the angle of scattering.

Questions

Q1. Since light behaves as a particle can you expect light to be affected by gravity? Explain or give reasons.

Q2. You can sail interstellar space with photons from the sun hitting a sail on a spaceship. Is it better to have a polished shiny surface or a black (absorbing) one for the sail? Why?

Q3. Two light bulbs emit yellow light. One is rated at 100 W and the other at 40 W. Which emits higher energy photons? Why?

Q4. If the temperature of black body radiation is doubled, by how much does the wavelength at the maximum of the power radiated change? Does the wavelength at the maximum get shorter, stay the same, or get longer?

Q5. In a Compton scattering off electrons, the electrons are scattered in the forward direction. What is the direction of the scattered photons? Why?

Q6. Compare the change in wavelength of a backward Compton scattered photon from electrons, protons, and a nitrogen atom.

Q7. In the photoelectric effect, does the current increase, decrease, or stay the same if the frequency of the photons is increased? What happens if the intensity increases? (The intensity of light is the energy per unit time per unit area perpendicular to a beam. It is determined by the number of photons in a beam). Explain.

Q8. In the photoelectric effect, what happens to the number of electrons emitted per unit time if the frequency of the photons is above the threshold and is increased? What happens to the maximum momentum of the electrons?

Q9. Does suntanning depend on the energy of the photons? Does it depend on the intensity of the photons? Explain.

Problems

P1. The threshold for the photoelectric emission of electrons from a certain material is a frequency of 1.2×10^{15} Hz. The frequency of light which shines on this material is 1.6×10^{15} Hz.
(a) Find the maximum kinetic energy of electrons emitted in eV and Joules.
(b) Determine the maximum velocity of the electrons.
(c) Find the stopping voltage.
(d) If the current obtained is 1.2 μA, what is the minimum number of photons that hit the plate per second?

P2. In a Compton scattering experiment off electrons in an atom, the incident light is in the X-ray region with an energy of 15.8 keV.
(a) Can the electrons be considered as (quasi-)free? Why? (See Chapter 20).
(b) Find the wavelength of forward scattered photons. DO NOT SOLVE, BUT SET IT UP.
(c) Find the wavelength of backward scattered photons.

P3. An incandescent light bulb is rated at 60 W. It is only 2.0% efficient.

Assume the light is all in the yellow region ($\lambda = 570$ nm). What is the number of photons emitted by the light bulb per second?

P4. Light of frequency 3.0×10^{15} Hz strikes a metal surface and emits electrons with $K_{max} = 6.22$ eV.
(a) What is the kinetic energy of the electrons in Joules?
(b) What is the work function of the metal?

P5. For what angle of Compton scattering does the maximum change in wavelength occur? What is the maximum change in wavelength of photons scattered from electrons at rest? What is the maximum recoil momentum of the electron?

P6. The surfaces of stars are not sharp. The radiation they emit is in equilibrium with the hot gases in the outer layer of the star. We can treat their radiation as black body radiation. The maximum of the power emitted by the sun occurs in the yellow region at 500 nm. What is the temperature of the sun's surface?

P7. Show that Planck's formula for the radiance of a blackbody at long wavelengths becomes

$$R(\lambda) = \frac{2\pi ckT}{\lambda^4}. \qquad (19.15)$$

This is a classical formula, called the Rayleigh–Jeans formula. Note that the Planck constant has disappeared in this limit.

P8. In a Thomson type of experiment, there is a vertical electric field of 100.0 V/cm and a horizontal magnetic field of 0.25 T.
(a) What is the velocity of the electrons along the x-axis to have no deflection?
(b) If the electric field is confined to a region of 0.50 cm, find the deflection of the electron beam from the x-axis. (The deflection is the angle that the electron beam would make with the horizontal axis after it leaves the electric field).

P9. Sunlight hits the earth at a wavelength of 500 nm (yellow) with an intensity of 665 W/m^2. How many yellow photons reach the earth per square meter per second? (See Q7 for the definition of intensity).

P10. If you double the velocity of electrons, by how much does their kinetic energy change? By how much does their Compton wavelength change? By how much does the energy of photons change when their momentum is doubled?

P11. (Challenge) In a Compton scattering experiment, the photon goes forward after the collision with an electron; it has lost 1/2 of its initial energy. Determine the momentum of the electron after the elastic collision, which conserves both energy and momentum.

Answers to odd-numbered questions

Q1. Yes, because energy and mass are equivalent. You can think of it as $hf = mc^2$.

Q3. Both emit the same energy photons.

Q5. To conserve energy and momentum, the photon must also be in the forward direction.

Q7. The current stays the same; if the intensity increases, so will the current.

Q9. Suntanning depends on both the energy of the photons and on their intensity. Ultraviolet is more effective than red and the intensity tells you the number of photons hitting you per unit time.

Answers to odd-numbered problems

P1. (a) $h = 4.135 \times 10^{-15}$ eV-s, $\Delta f = (1.6 - 1.2) \times 10^{15}$ Hz;
Maximum $KE = h\Delta f = 1.65$ eV $= 1.65 \times 1.6 \times 10^{-19} = 2.64$ J.

P3. $60\,\text{W} \times 0.02 = Nhf = N \times 6.63 \times 10^{-34}$ J-s$\times 3 \times 10^8$ m/s/(570×10^{-9} m),
$N = 3.4 \times 10^{18}$.

P5. Maximum change in λ occurs for backward electrons; $\lambda' = \lambda + 4.8 \times 10^{-12}$; $(\lambda' - \lambda) = 4.8 \times 10^{-12}$ m.

P7. Expand Eq. (19.2): $e^x \approx (1 + x)$ for small x. Thus
$I = \frac{2\pi c^2 h}{\lambda^5} \frac{\lambda k_B T}{hc} = \frac{2\pi c}{\lambda^4} k_B T$.

P9. 665 J/s-m$^2 = Nhf = N \times 6.634 \times 10^{-34}$ J-s$\times 3 \times 10^8$ m/s/(5×10^{-7}) m $=$
$N \times 3.98 \times 10^{-19}$ J. $N = 1.67 \times 10^{21}/$m^2-s.

P11. Momentum conservation: $h/\lambda' + p'_e = h/\lambda$; $0.5h/\lambda = p'_e$; also, we have $0.5hf = p'^2_e/2m$. Both momentum and energy conservation can only be satisfied if $p'_e c = p'^2_e/2m$ or $p'_e c = 2mc^2 = 1.02$ MeV.

Chapter 20

Atoms

20.1 The Hydrogen Atom

This book is organized by going from the most familiar to less familiar topics. We have now entered entirely unfamiliar territory and will explore it. Because electrons could be obtained from many atoms, it was assumed that all atoms contain electrons. Since atoms are neutral there must be positive charges to compensate for the electrons. The simplest atom is hydrogen, which was thought to be about 2000 times more massive than the electron. The early 1900s was a period of model building of atoms. How are the electrons and positive charges arranged? Perhaps, the most accepted model was the "pudding and raisin" model of J.J. Thomson. The pudding was the positive charge and the raisins were the electrons embedded in the pudding. The model turns out to be all wrong.

Ernest Rutherford was a New Zealand physicist who moved to Cambridge in England. He was a superb experimentalist. He thought of a way to probe the structure of atoms. He took alpha particles, which had been discovered a few years earlier in radioactive decays, and are now known to be the nuclei of He atoms to bombard a thin silver foil. If the plum pudding model was right, then the alpha particles should sail right through the silver, with almost no effect, since it is like sending a bowling ball through jello with a few embedded raisins (electrons). To his great surprise, a few alpha particles (about 1/800) were completely turned around! He concluded that the atom must contain a very concentrated charged mass at its center. From the path of the scattered alpha particles, he could tell that this *nucleus* must be smaller in extent than about 10^{-14} m. If the positive charge is in the center of the atom, then the alpha particle meets

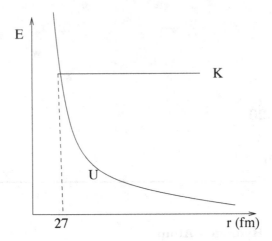

Fig. 20.1 Analogy to Rutherford's alpha particle scattering experiment. The initial kinetic and potential energies are shown.

a Coulomb barrier due to the repulsion of the positive charge of the alpha particle and that of the nucleus. The closer the alpha particles gets to the nucleus, the stronger this repulsion, as shown in Fig. 20.1. The Coulomb repulsive force is $kZze^2/r^2$, where ze is the charge of the alpha particle ($z = 2$ for He) and Ze that of the silver ($Z = 47$ for Ag). The potential energy of such a point or tiny spherical charge is $kzZe^2/r$ and the alpha can get no closer to the center than its kinetic energy allows (see Fig. 20.1). The zero of the potential energy is at an infinite distance from the atom. The distance of closest approach is reached when the kinetic energy has all been transformed to potential energy. This is called *the turning point*. It is very much like a ball with an initial kinetic energy that suffices to get it to roll part way up a hill, but not enough to reach the top.

We can find the distance of closest approach, when the potential energy is equal to the initial kinetic energy of the alpha particle, which was 5 MeV.

$$U = \frac{kZze^2}{R} = 2.17 \times 10^{-26} \text{ J-m}/R = K = 5 \text{ MeV} = 8 \times 10^{-13} \text{ J} \quad (20.1)$$

$$R = \frac{2.17 \times 10^{-26} \text{ J-m}}{8 \times 10^{-13} \text{ J}} = 2.7 \times 10^{-14} \text{ m}. \quad (20.2)$$

We have used $Z = 47$ for Ag and $z = 2$ for He nuclei. The nucleus of hydrogen with $Z = 1$ (a proton) must have a smaller radius. How are the electrons in a heavier nucleus distributed? What keeps the electron(s) from falling into the nucleus? And what can possibly keep the protons together in the tiny nucleus despite their Coulomb repulsion? We will answer these questions.

Niels Bohr took the Rutherford model of the atom: a tiny positively charged nucleus surrounded by electrons. He used the motion of the planets as an analogy. In both cases there is an attractive force proportional to $1/r^2$, GMm/r^2 for planetary motion and kZe^2/r^2 for the atom. Bohr made several assumptions:

(1) Classical mechanics fails at very small distances, like those in atoms.
(2) The electrons are in circular (rather than elliptic) orbits, but only in certain allowed *stationary* ones, in which no radiation occurs, so that the electrons do not fall into the nucleus.
(3) Orbital angular momentum is quantized,

$$L = mvr = n\hbar,\tag{20.3}$$

where $\hbar = h/(2\pi)$ is called *hbar* and occurs often. With Bohr's assumptions, we can now determine the radii and energies of allowed orbits for hydrogen-like atoms. (An example is doubly ionized Li^{++}, which has a single electron outside of a nucleus of charge 3e). Note that, in the first equation we have taken the Coulomb force, kq_1q_2/r^2, to provide the centripetal force.

$$F = = -kZe^2/r^2 = -mv^2/r,\tag{20.4}$$

$$v_n = \frac{kZe^2}{mvr} = \frac{kZe^2}{n\hbar} \le \frac{kZe^2}{\hbar} = \frac{cZ}{137},\tag{20.5}$$

$$r_n = \frac{n^2\hbar^2}{mkZe^2} = a_0n^2 = 5.29 \times 10^{-11}n^2/Zm,\tag{20.6}$$

$$E_n = -\frac{mk^2Z^2e^4}{2n^2\hbar^2} = -13.6Z^2eV/n^2.\tag{20.7}$$

For hydrogen $Z = 1$. In these formulas, we call $E = 0$ when the electron is far removed from the atom. Thus, a negative energy means that the electron is *bound*. The *quantum number n* is an angular momentum quantum number, the smallest one ($n = 1$) for hydrogen is called the Bohr radius, $a_0 = 5.29 \times 10-11$ m. You can think of the negative energy as the binding energy of the electron in the atom, e.g. like being tied to the nucleus by a spring; it takes energy to (break the bond) remove the electrons. The minimum energy needed occurs when the electron leaves the hydrogen atom with zero kinetic energy. This energy is called the *binding energy* or

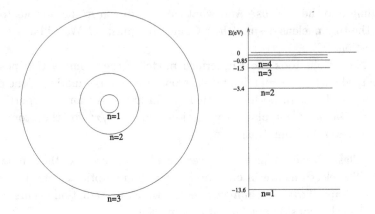

Fig. 20.2   Bohr orbits and energies for hydrogen.

ionization energy of the hydrogen atom. Since the potential energy is zero
at infinity, the minimum energy required to remove an electron in the nth
orbit is E_n. For the hydrogen atom in the ground state ($n = 1$, $Z = 1$), this
energy is 13.6 eV; the kinetic energy of the electrons is $(1/2)|U| = 13.6$ eV,
the energy needed to remove an electron; here U is the potential energy.
The relationship $K = 1/2|U| = -E$ holds for all n:

$$K_n = \frac{1}{2}mv^2 = \frac{1}{2}\frac{mk^2e^4}{n^2\hbar^2}, \tag{20.8}$$

$$U_n = \frac{-ke^2}{r} = -\frac{mk^2e^4}{n^2\hbar^2} = -2K, \tag{20.9}$$

$$E_n = U + K = -K. \tag{20.10}$$

For H-like atoms with $Z > 1$, this factor of Z must be included in all of the
above formulas, and e^2 becomes Ze^2.

Illustrative Problem 20.1: (a) For a H-like atom with $Z = 2$ (e.g., He$^+$),
find K, U, and E for the $n = 2$ orbit.
(b) What is the radius of the $n = 2$ orbit?
(c) What is the speed of the electron in this orbit?
(d) Find the f and λ when the electrons returns to its ground state.

Answers: (a) The energies are Z times the energies in the H atom. Thus
$E_2 = -2 \times 13.6$ eV/$n^2 = -13.6$ eV/2 $= -6.8$ eV; $K_n = 6.8$ eV,
$U_2 = -13.6$ eV.

(b) $r_2 = n^2/Za_0 = 2a_0 = 1.06 \times 10^{-10}$ m,

(c) $v = (Z/n)v_H = c/137$.

(d) $hf = E_2 - E_1 = -6.8$ eV $- (-27.2$ eV$) = 20.4$ eV, $\lambda = c/f = hc/hf = 1.97 \times 10-7$ eV-m$/20.4$ eV $= 9.66 \times 10^{-9}$ m.

Bohr was egged on by the observation some years earlier of a line spectrum for hydrogen in the visible region. Many people tried to fit a formula to the observed lines. Joe Balmer, a high school teacher, got the answer for the observed 4 lines in hydrogen

$$\lambda = \text{constant}\frac{4m^2}{m^2 - 4}, \tag{20.11}$$

with m an integer larger than 2. The constant, 3.65×10^{-7} m could be obtained from experiment. The frequencies are

$$f = \frac{c}{\lambda} = \text{constant}'\left(\frac{1}{4} - \frac{1}{m^2}\right). \tag{20.12}$$

Balmer went to see one of the professors at the ETH (a famous Swiss university in Zurich) and was told about the discrete lines observed in the radiation from stars. He then generalized his formula, and it was generalized further by Rydberg and Ritz in 1908 to

$$\frac{1}{\lambda} = R_H\left(\frac{1}{n^2} - \frac{1}{m^2}\right), \quad n < m, \tag{20.13}$$

with $n = 2$ for what is now known as the Balmer series. R_H is called the Rydberg constant for hydrogen, $R_H = 1.097 \times 10^7$ m^{-1}. The series for $n = 1$ is called the Lyman series and that for $n = 3$ the Paschen series. Bohr was able to derive the Rydberg constant. He assumed that the electron is excited from $n = 1$ to a higher energy level if it is given the right amount of energy. Once there, the electron will jump down to a lower level until it reaches the ground (lowest) state. In doing so, it emits radiation,

$$hf = E_i - E_f = E_0\left(-\frac{1}{n_i^2} + \frac{1}{n_f^2}\right), \tag{20.14}$$

$$\frac{1}{\lambda} = \frac{E_0}{\hbar c}\left(-\frac{1}{n_i^2} + \frac{1}{n_f^2}\right) = R_H\left(\frac{1}{n_f^2} - \frac{1}{n_i^2}\right). \tag{20.15}$$

The signs are correct because the energies are negative. Since R_H could be calculated from Bohr's formula, something had to be right. But Bohr could not predict when the jumps occurred or what path would be taken by the electrons (e.g., $n = 3$ to $n = 2$ to $n = 1$ or $n = 3$ to $n = 1$ directly). Nor could he predict the intensity of the various observed lines. When it came to other gases than hydrogen, e.g., helium, the Bohr formula did not work. It worked for lithium and a few other atoms. Moreover, in a magnetic field, the lines were found to split into two lines and this was not understood. Also, why can the angular momentum not be zero? Clearly, the Bohr model was inadequate for understanding other atoms than hydrogen.

20.2   The Structure of Atoms

In order to understand the structure of more massive atoms, we need to define the *exclusion principle*. Electrons are one of a class of particles called *Fermions*, named after Enrico Fermi. Wolfgang Pauli postulated that no more than one of these particles can be at the same place at the same time or can populate a given "state", labeled by its own quantum numbers. For an atom, these states differ from those of Bohr. What he called the angular momentum quantum number is now called the *principal quantum number*, labeled by n. Each such state can have various *orbital angular momenta* from $0 \ldots (n-1)\hbar$ (note that the angular momentum can be zero) and its projection on the z-axis is labeled by $m\hbar \leq \ell$, but m can be both positive and negative. Thus there are 3 quantum numbers versus Bohr's single one. It turns out that there can be two electrons in each of these states because electrons and other fermions can spin about their own axes (like the earth) and that spin is quantized as "up" or "down", depending on whether the spin is clockwise or counterclockwise. According to Pauli, we can only have 2 electrons in any state labeled by the quantum numbers $n\ell$, and m, one with spin up and the second one with spin down.

If excited to energies above the ground state, but below the ionization energy, an atom emits light, with a spectrum that is characteristic of that atom. Atoms can be identified by their spectra. For instance, for a hydrogen atom in the $n = 2$ state, the frequency of light emitted is $13.6 \times 3/4$ eV/$h =$ 10.2 eV/$h =$ 10.2 eV/4.14×10^{-15} eV-s $= 2.46 \times^{15}$ Hz.

Atoms have electrons in *shells*. A shell consists of a series of states with similar energies. The lowest shell has $n = 1\hbar$, $\ell = 0$, $m = 0$. Thus, it can contain only 2 electrons, one of spin up and one of spin down. The two atoms that make up this shell are H and He. The next shell has $n = 2\hbar$, $\ell = 1\hbar$, and $m = \pm 1\hbar, 0$ because m can run from $+l\hbar$ to $-l\hbar$. Thus this shell can have 6 electrons Li, Be, B, C, N, and O. This build-up of atoms continues. Atoms with closed shells are "noble" gases, e.g., He and Ne, and tend not to interact chemically with other elements. All atoms with one electron outside of a closed shell are good electrical and heat conductors, because the electrons tend to be loosely bound. It is for these atoms that Bohr's theory worked with an effective value of $Z_{eff} \leq Z$.

The light from stars, perhaps surprisingly, is just like that on the earth and is comprised of lines that correspond to those of hydrogen atoms. It is thus concluded that stars consist of a considerable amount of hydrogen gas.

Questions

Q20.1. Is it easier to ionize a hydrogen atom when the electron is in its ground state or an excited state? Explain.

Q20.2. In the Bohr model, sketch the deexcitation paths that an electron can take from the $n = 4$ state to $n = 1$.

Q20.3. As n increases, does the kinetic energy of the electron in H increase or decrease? Explain.

Q20.4. As n increases in the Bohr model, does the potential energy increase or decrease? What about the kinetic energy? Explain.

Q20.5. How many electrons can populate the $n = 3$, $\ell = 2$ state in H? Explain.

Q20.6. What is the least energy required to ionize (remove the electron) from the hydrogen atom in a state of quantum number n?

Q20.7. Why is the proton's motion not considered in a description of the H atom?

Q20.8. How much energy is required to ionize a H atom from the $n = 2$ state?

Problems

P20.1. What wavelength photon is required to ionize a hydrogen atom in its ground state?

P20.2. Rutherford showed that the nuclear radius of hydrogen was $r \leq$ 0.8×10^{-15} m. It is actually $\approx 10^{-16}$ m.
(a) What energy alpha particle (in eV) is needed to reach the nucleus?
(b) What fraction of space of the atom is occupied by the nucleus? Assume that the size of the atom is that of a sphere of radius $= a_0$, the Bohr radius.

P20.3. What energy photon is required to excite the hydrogen atom to the $n = 2$ state of hydrogen?

P20.4. What is the velocity of the electron in the $n = 2$ state of hydrogen?

P20.5. What levels are involved in the emission of a photon of energy $=$ 1.89 eV in hydrogen?

P20.6. The ionization energy for the outermost electron in Na is 5.1 eV. What Z_{eff} is required in the Bohr theory if the outermost electron is in an $n = 3$ state?

P20.7. Calculate the wavelength(s) of radiation emitted after an electron has made a transition from the ground state to the $n = 3$ state of H.

P20.8. A doubly ionized Li (Li^{++} atom has only one electron. Thus, Bohr's theory may be used. What are the radius and energy for $n = 3$ for Li^{++}?

P20.9. (Challenge) Recognizing that a photon has a momentum $p = E/c$, calculate the recoil velocity of a H atom, which is initially at rest, after it makes a transition from the first excited to the ground state.

P20.10. The energies of the ground state and first excited states of H are -13.6 and -3.4 eV, respectively. What is the frequency of radiation that can excite the H from its ground to first excited state? Neglect recoil.

Answers to odd-numbered questions

Q1. It is easier for a hydrogen atom in an excited state, since less energy is needed.

Q3. As n increases, the K.E. decreases; the velocity also decreases.

Q5. For $n = 3$, $\ell = 2$, there are $(2\ell+1)$ states. Since 2 electrons can inhabit any state, there $5 \times 2 = 10$ electrons at the maximum.

Q7. The proton has mass which is 1876 times that of the electron; its motion is thus much slower and its radius is much smaller than that of the electron.

Answers to odd-numbered problems

P1. $hf = 13.6$ eV; $\lambda = \frac{ch}{fh} = 12.4 \times 10^{-9}$ eV-m/13.6 eV $= 0.91$ nm.

P3. 13.6 eV $\times \frac{3}{4} = 10.2$ eV.

P5. $n = 2$ and 3.

P7. For $n = 3$ to 1: $13.6 \times (8/9) = 12.1$ eV;
for $n = 3$ to 2: $13.6(1/4 - 1/9) = 1.89$ eV;
for $n = 2$ to 1: 13.6 eV $\times 0.75 = 10.2$ eV.
The wavelengths are found from $\lambda = hc/hf = hc/E$.

P9. $\Delta E = 13.6 - 3.4 = 10.2$ eV if recoil is neglected. The momentum is then $10.2/(3 \times 10^8$ m/s$) = 3.4 \times 10^{-8}$ eV-s/m. To conserve momentum, the recoil momentum must be the negative of this value, and the velocity is p/M or pc/Mc and $v/c = pc/Mc^2 = 3.4$ eV/938 $\times 10^6$ eV $= 3.62 \times 10^{-9}$; $v = 1.08$ m/s.

Chapter 21

Particles and Waves — An Introduction to Quantum Mechanics

21.1 Matter Waves

Classically, particles and waves are easily distinguished. A stone is not a wave but it can make water waves if it is dropped into a small pond. Two stones can't be in the same place at the same time; two waves can do so and interfere with each other. In the last chapter, it is was therefore unexpected that light can behave as either a wave or a particle.

If light can behave as a particle, why can't particles behave as waves? What would be their wavelengths? For light $E = hf$ and $\lambda = v/f = hv/E$. For particles with velocity v, the kinetic energy is $E = (1/2)mv^2$ if m is the mass of the particle; then $\lambda = h\sqrt{\frac{2}{mE}}$. This looks complicated. For light, the momentum p is

$$p = \frac{E}{c} = \frac{hf}{c} = \frac{h}{\lambda}. \qquad (21.1)$$

This looks more reasonable; there are no square roots. Louis de Broglie in France pondered both questions and came to the conclusion that, if particles can behave as waves, then Eq. (1) should apply, or

$$\lambda = h/p, \qquad (21.2)$$

which is now called the *de Broglie wavelength* of a particle. A massive particle will have a minutesimal wavelength, so that the wave properties would not be observed. As an example, a 1 g mass with a speed of 1 m/s would have a wavelength of $\approx 6.6 \times 10^{-28}$ m. This is too small to observe. However, if you take the lightest particle known at that time, the electron with a mass of 9.1×10^{-31} kg, then at $v = 1$ m/s, $\lambda \approx 7 \times 10^{-3}$ m, and even with $v = 10^6$ m/s, $\lambda \approx 7 \times 10^{-9}$ m, and we are then in a region where interference effects could be made visible. Note that for light $E = pc = hc/\lambda$, but for particles $E = \frac{p^2}{2m} = \frac{h^2}{2\lambda^2 m} \neq pc$.

Illustrative Problem 20.1: Find the energy for light and an electron with a wavelength of 10^{-10} m.

Solution: The energy for light is $hf = hc/\lambda = 12.4 \times 10^{-9}$ eV-m/10^{-10} m $= 124$ eV $\approx 1.2 \times 10^2$ eV. For an electron, the momentum is $h/\lambda = 6.63 \times 10^{-34}$ J-s/10^{-10} m $= 6.63 \times 10^{-24}$ kg-m/s. The kinetic energy is $p^2/2m = (6.63 \times 10^{-24}$ kg-m/s$)^2/(2 \times 9.11 \times 10^{-31}$ kg$) = 2.41 \times 10^{-17}$ J $= 1.5 \times 10^2$ eV, which is not the same as for light!

21.2   Electron Waves — Diffraction

If electrons or other particles can behave as waves, we should be able to observe interference effects. Clinton Davisson and Leston Germer figured out that diffraction should be observable by shining electrons on a crystal, since the wavelength can be made akin to X-rays. Crystals have regular arrays of atoms with a spacing of the order of 10^{-10} m. Davisson and Germer showed that the diffraction of electrons was similar to that of X-ray waves; see Fig. 18.5. This experiment thus showed that electrons can diffract from crystals, and can, indeed, behave as waves. Their wavelength was shown to be given by Eq. (21.1).

We see that not only can light behave as particles (photons), but particles can also behave as waves. This is, in fact, the basis for electron microscopes. The advantage of electron microscopes over ordinary ones is that electrons can be accelerated to a relatively high energy, so that the wavelength can be made small. For instance for a 10 keV electron, $pc \approx 10^5$ eV, and $\lambda = h/p = hc/pc \approx 4.1 \times 10^{-15} \times 3 \times 10^8$ eV-m/10^5 eV $\approx 1.2 \times 10^{-11}$ m. We have used $hc = 12.3 \times 10^{-7}$ eV-m. Diffraction has taught us that shorter wavelengths allow you to see more details.

It is not only electrons that can behave as waves, but all material particles such as hydrogen atoms. Neutrons, the neutral partners of protons, are used to investigate details of crystal structures, which is helped by their short wavelengths. Their mass is about 1800 times that of electrons.

21.3   The Uncertainty Principle

In classical physics, we can make measurements to arbitrary precision; this is no longer true in quantum mechanics, which applies at very short distances. Consider diffraction from a single slit. The first minimum occurs when $\sin\theta = \lambda/a$, where a is the size of the opening. If $\lambda \approx a$, it may just be possible to get to the first minimum, since $\sin\theta \approx 1$ there. The difference in path lengths between the interfering "beams" is $\lambda/2 = h/(2p)$. Since the transmitted photons can be anywhere between the central maximum and first minimum, we have

$$\Delta x \Delta p \sim h/2. \tag{21.3}$$

We have used optics but also the de Broglie wavelength to get this result. It is the basis of Werner Heisenberg's famous *uncertainty principle*,

$$\Delta x \Delta p \geq \hbar. \tag{21.4}$$

This is an absolute limit. It says that you cannot measure or know the position and momentum of anything at the same time to arbitrary precision. You can see why, physically: To observe a particle you have to shine light on it, but light has momentum and thus imparts some of it to the particle. If the particle was at rest to begin with, it does not remain at rest. If you use very low frequency light, the wavelength gets very large and diffraction effects spoil the precision. You cannot localize the particle to roughly better than one wavelength. This is not relevant, in a practical sense for massive objects, since the wavelength is then very short.

There is also an uncertainty relation for energy and time

$$\Delta E \Delta t \geq \hbar. \tag{21.5}$$

This means that you need infinite time to measure the energy accurately. This is only possible for the ground state of atoms and other objects because excited states decay and don't live forever.

Illustrative Problem 21.2: The lifetime of the $n = 2$ state of hydrogen is 1.2×10^{-15} s (not actual lifetime). Find the natural line width of the $n = 2$ state.

Solution: The width of the state is $\Delta E = \hbar/\Delta t = 6.58 \times 10^{-16}$ eV-s$/(1.2 \times 10^{-15})$ s $= 0.55$ eV.

21.4   Introduction to Quantum Mechanics

Quantum mechanics, which replaces classical physics at short distances, is based on many of the above ideas.

It is a wave theory for particles. The waves are probability waves. The electrons in a hydrogen atom, for instance, do not have definite circular paths. Instead there is a probability of finding the electrons at some distance from the nucleus. This probability is a maximum at about the Bohr radius. The atomic orbits are standing probability waves in three dimensions.

The de Broglie wavelength of an electron also provides a reason for the allowed Bohr orbits,

$$L = n\hbar = nh/(2\pi) = mvr = pr = hr/\lambda, \qquad (21.6)$$

$$2\pi r = n\lambda. \qquad (21.7)$$

The Bohr orbits are thus those which can fit a whole number of standing de Broglie waves around the circumference.

Consider an electron contained in a box the size of the hydrogen atom, for which we take the Bohr radius, i.e., the side of the box is $L = 5.3 \times 10^{-11}$ m. To get a standing wave, we expect the wave to vanish at the walls of the box. The maximum wavelength occurs if these are the only nodes, so that the $L = \lambda/2$. We then find that the momentum of the electron in the box is $p = h/\lambda = 6.63 \times 10^{-34}$ J-s$/10.6 \times 10^{-11}$ m $= 6.25 \times 10^{-24}$ kg-m/s and the velocity of the electron is $p/m = 6.25 \times 10^{-24}$ kg-m/s$/9.11 \times 10^{-31}$ kg $= 6.86 \times 10^{6}$ m/s. This is approximately the speed we found for the electron in its lowest orbit. The minimum momentum is the uncertainty of the momentum and, since the particle can be anywhere in the box, we also regain the Heisenberg uncertainty principle, i.e., $\Delta p_x \Delta x \approx p\Delta x = 6.25 \times 10^{-24}$ kg-m/s $\times 5.3 \times 10^{-11}$ m $= 3.3 \times 10^{-34}$ J-s $\approx h/2$.

Questions

Q21.1. Suppose that $\hbar$ is large, $\hbar = 1$ J-s. If you throw a ball to your friend, what is the approximate chance of hitting her? Take a 10 g ball and $v = 10$ m/s.

Q21.2. An electron is confined within a space of 1.5×10^{-10} m. Is it possible for the electron to be at rest? If not what is its minimum velocity?

Q21.3. In a hydrogen atom, can an electron be localized exactly? Explain.

Q21.4. An electron is bound in an atom of radius 1.2×10^{-9} m and another one is bound in a nucleus of radius 1.2×10^{-15} m. Find the ratio of their minimum momenta and kinetic energies.

Problems

P21.1. What is the wavelength of a neutron of kinetic energy $= 0.1$ eV? The mass of a neutron is 1.68×10^{-27} kg.

P21.2. A proton and an electron are both confined in a one dimensional "box" the size of the nucleus (1.5×10^{-15}) m. What are their minimum momenta, kinetic energies and velocities?

P21.3. Calculate the de Broglie wavelength of a 1 eV electron.

P21.4. The central maximum width in a diffraction experiment can be taken as reaching to the first minimum. If it is the same for blue light $(\lambda = 460$ nm$)$ and electrons, find the ratios $K_{el}/K\gamma$, $v_{el}/v\gamma$, and $p_{el}/p\gamma$ $(\gamma$ is light$)$. The distance to the screen is assumed to be large.

P21.5. A molecule of nitrogen (mass $= 14$ times that of hydrogen) is trapped in a blood vessel of diameter 0.26 mm. What is the uncertainty of its velocity?

P21.6. An electron of velocity 3×10^6 m/s has what wavelength? If a photon has the same momentum, find the wavelength of the photon.

P21.7. A certain level of hydrogen decays to a lower level in 2.4×10^{-15} s. What is the uncertainty (or width) of its energy level in eV?

P21.8. Find the wavelength of a photon, electron, and proton of kinetic energy $= 1$ keV.

P21.9. The de Broglie wavelength of protons is 1.45×10^{-14} m. What is the kinetic energy of these protons?

P21.10. If you double the velocity of electrons, by how much does their de Broglie wavelengths change?

P21.11. Calculate the wavelengths of radiation emitted after an electron has made a transition from the ground state to the $n = 3$ state of H.

P21.12. If the momentum of a photon is $p = E/c$, calculate the recoil velocity of a H atom which is initially at rest, after it makes a transition from the first excited state ($N = 2$) to the ground state.

P21.13. Neutrons are diffracted by a single slit. The velocity of the neutrons is 3.3×10^3 m/s. The slit is circular with a radius of 0.14 mm. For what angle will the first minimum occur in the diffraction pattern?

Answers to odd-numbered questions

Q1. $\Delta x \Delta p \geq \hbar$; thus $\Delta x \geq 1$ J-s$/(10^{-2}$ kg $\times 10$ m/s$) = 10$ m. The chances of hitting your friend is not very high.

Q3. No, since $\Delta x \geq \hbar/\Delta p \geq \hbar/p$. Since $p \neq \infty$, $\Delta x \neq 0$.

Answers to odd-numbered problems

P1. $\lambda = h/p = h/\sqrt{2mK} = 6.63 \times 10^{-34}$ J-s$/$
$\sqrt{2 \times 1.68 \times 10^{-27}}$ kg $\times 10^{-1}$ eV$/1.6 \times 10^{-19}$ eV/J $= 1.45 \times 10^{-24}$ m.

P3. $\lambda = h/p = hc/(pc) = hc/\sqrt{2mc^2K} = 1.23 \times 10^{-11}$ m, since $mc^2 = 0.51$ MeV for an electron.

P5. $\Delta p \geq \hbar/\Delta x = 1.055 \times 10^{-34}$ J-s$/(0.26 \times 10^{-3}$ m$) = 4.06 \times 10^{-31}$ kg-m/s; $\Delta v = \Delta p/m = 1.73 \times 10^{-5}$ m/s.

P7. $\Delta E \geq \hbar/\Delta t = 6.58 \times 10^{16}$ eV-s$/(2.4 \times 10^{-15}$ s$) = 0.27$ eV.

P9. $K = p^2/(2m) = h^2/(\lambda^2 2m) = (hc)^2/\lambda^2 2mc^2)$. Now $hc = 1239$ MeV-fm, where 1 fm $= 10^{-15}$ m. Thus $K = (1239$ MeV-fm$)^2/((14.5$ fm$)^2 \times 2 \times 938$ MeV$) = 3.89$ MeV because mc^2 for the proton is 938 MeV.

P11. For $n = 3$ to 1: $13.6 \times (8/9) = 12.1$ eV; for $n = 3$ to 2: 13.6 eV $(1/4 - 1/9) = 1.89$ eV; for $n = 2$ to $n = 1$: 13.6 eV $\times 0.75 = 10.2$ eV. The wavelengths are found from $\lambda = hc/hf = hc/E$.

P13. $\sin \theta = \lambda/d = h/(pd) = hc/(pcd)$; $d = 2.8 \times 10-4$ m; $pc = mc^2v/c = 9.39 \times 10^8$ eV $\times (3.3 \times 10^3/3 \times 10^8)$. Thus $\sin \theta \approx \theta = 4.31 \times 10^{-7}$.

Chapter 22

The Special Theory of Relativity

"When a man sits with a pretty girl for an hour, it seems like a minute. But let him sit on a hot stove for a minute — it's longer than an hour. That's relativity."... Albert Einstein

We have seen that Newtonian mechanics breaks down at very small distances. It also fails at very large velocities, namely those approaching the speed of light. The mechanics at very high speeds is the subject of this chapter.

22.1 Relative Motion

The relative motion of objects was studied in classical or Newtonian mechanics. An example is a boat on a river. If the velocity of the boat relative to the water is $\vec{v}_{boat}$ and that of the river is $\vec{v}_{river}$, then the velocity of the boat relative to the shore is $\vec{v}_{shore} = \vec{v}_{boat} + \vec{v}_{river}$. If the boat goes upstream $v_{shore} = v_{boat} - v_{river}$. The same situation holds for a passenger walking in a train. If we apply the boat example to sound waves, it works fine. But it no longer works for light waves.

If you do an experiment on a train moving at constant velocity, the results will be the same as those you would obtain on the platform. For instance, a ball thrown straight up comes straight down. A person standing on a stationary platform would see the ball move in a parabolic path, since the train moves during the time the ball is in the air. But Newton's laws are the same in any frame of reference moving at a constant velocity, called an *inertial frame* of reference. If there are curtains on a train moving at constant velocity, no experiment can be done which tells you that you are in motion.

Fig. 22.1   Albert Einstein at the time of "The Miracle Year", 1905. Picture courtesy of the American Institute of Physics, the Emilio Segrè Archives.

This is the first of Einstein's assumptions for his special theory of relativity:

I. *The laws of physics are identical in all inertial reference frames.*

22.2   Special Relativity

The second assumption Einstein made for his theory was that the velocity of light in vacuum is the same for any observer in any inertial frame. This differs markedly from our everyday experience. It says that the velocity of light in vacuum is c, whether you are at rest or moving at any constant velocity, v, even $v = c$.

In the early part of the 20th century it was still believed that light needed a medium for its propagation, and the medium was called an *aether*. The speed of light, like the boat's was relative to this aether. Albert Michelson and Edward Morley designed an experiment to detect the motion of the earth through this aether. At some point in the earth's motion around the sun, we should be moving in the same direction as the aether and six months later we should move in the opposite direction. Michelson and Morley used light and interference to detect this motion. The basic set-up is shown in Fig. 22.2. The mirror in the center is a "half-silvered mirror",

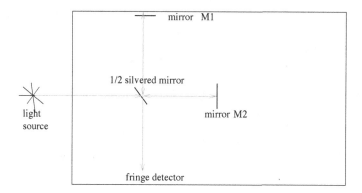

Fig. 22.2   Sketch of the Michelson–Morley experiment.

which both reflects and transmits light. The light then travels to mirrors M1 and M2. The detector, or eye, receives light from both M1 and M2. If the difference in optical path lengths for the reflection from mirrors 1 and 2 is $\lambda/2$, no light should reach the eye, but if it is a whole number of wavelengths they should see bright light. Michelson and Morley saw no variation in the brightness of the light throughout the year. The experiment is the basis of Einstein's second assumption. (It is not clear that Einstein was aware of the experiment and its results).

    II. *The speed of light in vacuum is the same in any inertial frame, regardless of the motion of the source or the observer.*

    If you travel at a speed of $v = c/2$, light will pass you at a speed of c. This is clearly quite different than the example of the boat on a river and has very large implications, some of which we shall now examine.

22.3   The Relativity of Time

To pursue the consequences of his two assumptions, Einstein "performed" some thought experiments. The consequences are very small unless velocities approach those of light. He thus considered a spaceship, as shown in Fig. 22.3. An astronaut in the spaceship sends a light signal from the ground to the ceiling of the space ship, a distance d away. A mirror is placed on the ceiling and the time elapsed between the sending of the light

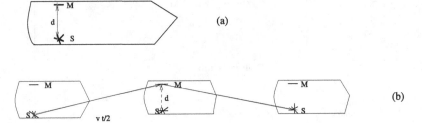

Fig. 22.3 Einstein's thought experiment with light beams (a) view inside spaceship; (b) view from earth.

and its return is measured. This time is

$$t_0^s = 2d/c, \tag{22.1}$$

where the superscript s stands for spaceship. For an observer on the earth, the light travels further, since the spaceship has moved between the time the light was sent and received again, as shown in Fig. 22.3(b). During the time the light travels from the floor to the mirror, the spaceship has traversed a distance $vt/2$, if v is the velocity of the space ship and t is the time for the light to reach the mirror and return. (Note that $t \neq t_0^s$).

Both the astronaut on the space ship and the individual on the earth will agree on the velocity, v, but the spaceship astronaut will see it as the velocity of the earth under her. The vertical distance d is unaffected by the motion of the space ship. The distance for half the flight of the light beam is then obtained by the Pythagorean theorem as

$$ct/2 = \sqrt{d^2 + v^2 t^2/4}, \tag{22.2}$$

where t is the time measured by the observer on the earth. Squaring both sides, we get

$$c^2 t^2/4 = d^2 + v^2 t^2/4, \tag{22.3}$$

$$(c^2 t^2/4)(1 - v^2/c^2) = d^2, \tag{22.4}$$

$$ct/2 = \frac{d}{\sqrt{1 - v^2/c^2}}, \tag{22.5}$$

$$t = \frac{2d/c}{\sqrt{1 - v^2/c^2}} = \frac{t_0^s}{\sqrt{1 - v^2/c^2}}. \tag{22.6}$$

The factor $\sqrt{1 - v^2/c^2}$ appears over and over again and is given a special symbol

$$\gamma = \frac{1}{\sqrt{1 - v^2/c^2}} . \tag{22.7}$$

The time for the observer on earth is longer by the factor γ than it is for the stationary observer on the space ship. The astronaut will conclude that the clocks on earth are slow. The time elapsed on the spaceship t_0^s is called the *proper time* and is the shortest time measured for the experiment. We use the subscript 0 for the proper measurement. The elapsed time is longer for the observer on earth (the moving observer) because light had to travel further due to the motion of the spaceship. Note that the observer at rest on the spaceship could do with one watch for both the start and end of the flight of light. The observer(s) on earth need two watches since the spaceship has moved (e.g., from New York to Chicago) during the experiment. This strange dilation of time occurs because the velocity of light is the same for the two observers, the one on the spaceship and the one on the earth.

We can check time dilation experimentally. It is difficult to accelerate an object to a speed close to that of light, but we can do so with very light particles, such as an electron. Certain radioactive particles have a short lifetime. One such particle, the muon (a heavy electron of mass about 210 times that of the electron) has a lifetime of only 2 μs. Suppose that a muon is produced at a speed of 0.6c. The lifetime you would observe if you could sit on the muon is 2 μs, because you are at rest relative to the muon. The lifetime in the laboratory, on the other hand, is γ times longer,

$$\gamma = \frac{1}{\sqrt{1 - v^2/c^2}} = \frac{1}{\sqrt{1 - 0.36}} = \frac{1}{\sqrt{0.64}} = 1/0.8 = 1.25 . \tag{22.8}$$

Thus, the measured lifetime in the laboratory would be 1.25×2 μs $= 2.5$ μs, and this is what is found. The experiment has been done for a muon traveling in a circle in a magnetic field.

22.4   The Contraction of Lengths

Since time is dilated for moving observers, do lengths also change? Yes. Let's examine this. On earth, we can use a long tape measure and mark off the distance that the spaceship has moved during the experiment. Let us call the length L_0. This distance is

$$L_0 = vt . \tag{22.9}$$

since the spaceship travels at speed v, and the time for the signal to return is t. This may be the distance from New York to Chicago. In the spaceship, the astronaut cannot measure this distance directly; that is, he cannot lay down a tape measure. The astronaut sees the earth move under her in the opposite direction from what the observer on earth sees for the spaceship, Since the earth is moving at a speed v, and the time elapsed for the experiment is t_0^s, the distance moved (NY to Chicago) is presumed to be vt_0^s, or

$$L^s = vt_0^s = vt/\gamma = L_0/\gamma < L_0. \qquad (22.10)$$

Thus, lengths parallel to the motion are foreshortened in a moving frame of reference. In this case, the astronaut is the moving observer because she cannot lay down a tape measure. Let's look at the muon again. What is the distance traveled by the muon during its lifetime? In the laboratory it is $vt = 0.6 \times 3 \times 10^8$ m/s $\times 2.5 \times 10^{-6}$ s $= 450$ m. The astronaut claims that this distance is $vt_0^s = 0.6 \times 3 \times 10^8$ m/s $\times 2 \times 10^{-6}$ s $= 360$ m $= 450$ m/γ.

You may have noticed that relativistic effects are of order v^2/c^2, so that, even for $v = 0.1c$, the corrections are only of the order of 1% rather than 10%. In Newtonian mechanics, when a ball is thrown straight up on a train moving at constant velocity, it also comes straight down again, because the laws of motion are unchanged in a frame moving at constant velocity (an inertial frame). An observer on a train platform would see the ball as going in a parabola, because the horizontal motion of the train has to be added to the verical motion of the ball. However, the height reached by the ball would be the same for the observer on the platform as the one on the train. The same is true in relativity. Distances perpendicular to the motion are unaffected by relativity. This is why Einstein could do his mirror experiment on the spaceship. Thus, distances perpendicular to the velocity $\vec{v}$ are unaffected by the frame of reference, just as in Newtonian mechanics.

22.5   The Addition of Velocities, Momentum, Energy

When the velocities of objects get to be close to those of light, velocities in different moving frames of reference don't just add or subtract, as they do at slow speeds. This is because nothing can go faster than light itself. If an object has a velocity v_1, in a frame moving at velocity v, then the velocity in a frame at rest would be

$$v_1' = \frac{v_1 + v}{1 + v_1 v/c^2} \,. \tag{22.11}$$

If either v_1 or $v = c$, then $v_1' = c$; even if both v_1 and v are $= c$, it follows that $v_1' = c$. This is in accord with the principle that the velocity of light is the same in any inertial frame.

Is momentum conserved when there are no external forces acting? It turns out that if momentum is defined as $\vec{p} = m\vec{v}$, then momentum may be conserved in one frame of reference, but not in all inertial frames. In order for momentum to be conserved in all inertial frames, we need to change the definition to

$$\vec{p} = \gamma m_0 \vec{v} \,. \tag{22.12}$$

We have here called the rest mass m_0, the mass of an object when it is at rest. What about kinetic energy? Earlier, we saw that the kinetic energy is equal to the work done on a body, Fd, if the force is in the direction of motion. The force is related to the momentum by $F = \frac{\Delta \vec{p}}{t}$. These are the ingredients. With them, one finds that

$$\text{K.E.} = \gamma m_0 c^2 - m_0 c^2 = (\gamma - 1) m_0 c^2 \,. \tag{22.13}$$

The first term, proportional to γ depends on the velocity of the particle; the second one does not. It is called the *rest energy*. We have used the subscript 0 for the *rest mass* of the object. As the speed of an object approaches c, its mass or inertia — for a change of motion — grows rapidly, so that it gets more and more difficult to accelerate the object to a still higher velocity. Indeed, gamma approaches infinity as v approaches c, so that it can never equal c. For small γ, that is for non-relativistic speeds,

$$\gamma = \frac{1}{\sqrt{1 - v^2/c^2}} \approx 1 + \frac{1}{2} v^2/c^2 \,, \tag{22.14}$$

so that K.E. becomes

$$m_0 \left(1 + \frac{1}{2} v^2/c^2\right) c^2 - m_0 c^2 = \frac{1}{2} m_0 v^2 \,. \tag{22.15}$$

We regain the Newtonian equation for the kinetic energy, as we should.

The total energy of an object is its rest energy + its kinetic energy,

$$E = K + m_0 c^2 = \gamma m_0 c^2 - m_0 c^2 + m_0 c^2 = \gamma m_0 c^2 \,. \tag{22.16}$$

For an object moving at velocity v, the mass becomes m

$$m = \gamma m_0 \,, \tag{22.17}$$

so that Eq. (22.17) can be written as $E = mc^2$. This is the famous Einstein equation, but m is the moving mass and not the rest mass. For relativistic motion the mass m is not constant, but increases with speed. As the velocity of the object approaches c, its mass approaches infinity. This is what makes it impossible to accelerate an object to a speed larger than c. We see that m is more than inertia for a change of motion; it is also a form of energy and changes with speed. In brief, mass is equivalent to energy ($= E/c^2$) Any time we do work on an object we change its energy and thus its mass. For example, by lifting the object through a height h, we change its potential energy and therefore its mass! The change in mass may be small, but it is not zero.

If we expand Eq. (22.17) we can show that the energy can be written in an alternative form,

$$E = \sqrt{p^2c^2 + m_0^2c^4}. \tag{22.18}$$

In this form, we see that for a particle of zero rest mass, $E = pc$. This applies to a photon, which travels at velocity c in any frame of reference. We thus have

$$p(photon) = E(photon)/c = hf/c. \tag{22.19}$$

If we write $m = \gamma m_0$, it appears that if $m_0 = 0$ then m is also zero, but this is not correct; for a particle of zero rest mass (e.g., a photon) $\gamma = \infty$ and $\infty \times 0$ is undefined.

Illustrative Problem 22.1: Find the velocity, momentum and total energy of an electron and a photon of kinetic energy = 0.4 MeV.

Answers: For an electron, the energy is $m_0c^2 + K = 0.51$ MeV $+ 0.4$ MeV $= 0.91$ MeV; the momentum is $\sqrt{E^2 - m_0^2c^4}/c = 0.75$ MeV/c; the velocity is found from $E = \gamma m_0 c^2, p = \gamma m_0 v$; thus $v/c = pc/E = 0.75/0.91 = 0.82$. For a photon of energy 0.4 Mev, $E = 0.4$ Mev, $pc = 0.4$ MeV, $v = c$.

Illustrative Problem 22.2: In a certain frame of reference, the velocity of an object is $v_1 = c/2$. The frame itself is moving at velocity $u = c/4$ in the opposite direction to v_1 relative to an observer on earth. What velocity will be measured by the earth observer?

Answer: The velocity observed on earth is $\frac{v_1 - u}{1 - v_1 u/c^2} = 2c/7$.

Time dilation applies also to heartbeats, pulse, and other physiological processes. In a moving frame, these processes are slowed down. This is the basis for one of the more famous puzzles of relativity, the so-called twin paradox. It goes as follows. There are two identical twins on the earth. One, A, leaves on a long trip to a distant planet in a spaceship that is able to travel at a speed close to that of light. The twin returns after a quick turn around at the distant planet. When she lands on earth, the twins compare their ages and twin B finds that she has aged more (say, 10 years) than twin A (say, 8 years). How is it possible for twin A to have aged less than twin B? Why is the situation not symmetrical, since A sees the earth recede at the same speed as B sees A travel to the distant planet? Assume that the turn-around time for twin A can be neglected. The situation should thus be symmetrical and twin A should see a younger twin B on her return. But that is hardly possible. They cannot differ in their answers. So, have they aged the same or is there a difference? If there is a difference, where does the asymmetry come from? You can also think about the problems as follow. You are standing on a platform when twin A passes you in a very rapid train. Both twins conclude that the other's watch is running slow. Twin A thinks twin B's is slow and vice versa. Twin A gets off at the next station and immediately catches a train to return. When she gets back, she is very surprised to find that her watch is slow compared to B, even though she thought B's was slow compared to hers on the way to the distant station. That is, twin A has aged less.

There is no real paradox if the situation is analyzed carefully. The answer to the paradox is that twin A has to change her reference frame from one traveling away from the earth (or to the next station) to one traveling towards it (or returning). This introduces the asymmetry and A will have aged less than B, because the heartbeat of the moving twin (A) during travel is slow compared to the twin who remains in a single inertial frame (B). The result can be checked experimentally. Clocks are now so accurate that the result can be checked for a satellite.

There are more effects than we can discuss here. For instance, two observers moving relative to each other will not agree on whether their clocks are synchronized. To see how this comes about, consider the astronaut in the spaceship. She can synchronize two clocks at either end of the ship by standing in the middle and sending a light signal simultaneously in both directions. Clocks on the wall at either end start when the light signal hits them. To the astronaut, these clocks are thus synchronous, but an observer on earth will not agree. If the observer is right under the astronaut as she sends the signal, then the rear of the spaceship is moving towards the source of the light, whereas the front is receding from it. Thus, for the earth observer, it takes less time for the light to reach the rear clock than the front clock. Thus, the person on earth will not see the two clocks as starting at the same time.

22.6 Introduction to General Relativity — The Principle of Equivalence

General relativity deals with accelerating frames of reference, and thus with gravity. We have already observed some phenomena in accelerating frames, as in a freely falling elevator. Let us examine this situation again, but let's assume an elevator accelerating upwards, as shown in Fig. 22.3. A person on a scale on the floor of the elevator will measure an apparent weight larger than mg, since the normal force, N, on the individual is $N = mg + ma$; the unbalanced force is $N - mg$. It appears that the gravitational constant g has increased to $g + a$. Einstein asked himself whether one can distinguish between a change of g and an acceleration. He concluded that you cannot do so. This is called the *principle of equivalence*:

You cannot tell the difference between an accelerating frame of reference and gravity.

Perhaps, then, gravity is nothing but being in an accelerated frame of reference.

What if you throw a ball horizontally in the elevator that is accelerating upward? While the ball is in the air, the elevator floor is moving up towards it, due to the acceleration, a. The path will thus not be horizontal, but will curve towards the floor of the elevator. The path would be the same in the gravitational field of the earth (i.e., the ball would fall to the earth), and we would blame it on gravity.

As another example, let's consider a beam of light. If the elevator is at rest in outer space, the path of a beam of light aimed horizontally at the far wall would follow a horizontal path. If the elevator accelerates upwards, then the light beam would be deflected towards the floor just like the ball in the previous example. The reason is that the floor is moving up while the light beam is in the air. The principle of equivalence then tells you that light is deflected by gravity. In a way, you can see this from $E = hf = m_{equiv}c^2$. The prediction was first checked experimentally during a total eclipse of the sun, because the effect is very small.

Fig. 22.4 Person in an elevator accelerating upwards.

Nowadays the bending of light in a gravitational field is used to focus light from distant stars as they pass other heavenly bodies, e.g., the sun. The effect is a gravitational lens.

Questions

Q22.1. Can we use $\vec{p} = \gamma m\vec{v}$ for $v \ll c$? Why?

Q22.2. If you compress a spring will its mass change? Explain.

Q22.3. If the velocity of an object approaches zero, does its energy also do so? Explain.

Q22.4. Can you tell whether an object is accelerating due to a force or just being close to a large mass? Explain.

Q22.5. Will a clock on the sun read the same time differences as one on earth? Explain.

Q22.6. A golf ball is driven down a teeway. Who measures the proper length, a person on the earth or an astronaut in a spaceship? Why?

Q22.7. In the twin paradox, explain why the elapsed time for twin A is less than for twin B, even though A travels at high speed.

Q22.8. A chess game is taking place in Seattle and is observed by A, who is in a spaceship moving at high speed, and also by B, who looks over the shoulders of one of the players. Will A or B measure the longer time between moves? Why?

Q22.9. A radioactive isotope with a 1/2 life (the time during which half the particles have decayed or died) $\tau_{1/2}$ is moving at high speed in a particle accelerator. Does the observer at rest in the laboratory measure the proper time for the 1/2 life? Explain.

Q22.10. A spaceship is moving at high velocity past B on earth. A is on board the spaceship. Does A or B measure a longer length for the spaceship? Explain.

Problems

P22.1. Determine the ratio $\frac{\Delta m}{m}$ when the mass m is carried to the top of Mt. Rainier, $h = 4500$ m.

P22.2. A spacecraft travels at a velocity of $0.95c$ relative to the earth. It is 82 m long, as measured by the astronaut on board.
(a) What is the length of the spacecraft, as measured by an observer on the earth?
(b) How much time does it take for the spacecraft to pass the observer on earth?
(c) On the spacecraft, how much time does it take for the earth observer to pass from the front to the back of the spacecraft?
(d) Use your answer to part (c) to obtain the length of the spacecraft for the astronaut.

P22.3. In the twin paradox, determine the aging of twin B if twin A has aged by 4 years. The spaceship travels at $0.8c$.

P22.4. An astronaut on a spacecraft measures the size of the ship: length = 77 m, diameter = 14 m. If the spaceship is traveling at $0.9c$ relative to the earth, what will be the dimensions of the spacecraft as measured on earth?

P22.5. A spaceship is traveling at $0.8c$ relative to the earth. The distance from New York to a midwestern city is 1500 km. What length would be measured on the spaceship?

P22.6. The sun radiates energy at a rate of 3.92×10^{26} J/s.
(a) What is the change of mass of the sun/s?
(b) What is the change of mass of the sun in 100 y?

P22.7. A spacecraft is traveling towards the moon. Its speed relative to the moon is $0.55c$. When it gets close, a moon explorer is launched from the craft and leaves it at a velocity of $0.55c$, as measured on the spaceship. What is the velocity of the moon explorer relative to the moon?

P22.8. Two particles of mass 6.5×10^{-23} kg approach each other on the way to a head-on collision. Their speeds are both measured at 2.1×10^8 m/s in the laboratory. What is the speed of one of these particles as seen by the other one?

P22.9. An electron is accelerated from rest by a voltage difference of 3.6×10^7 V. Determine the final velocity, momentum and kinetic energy of the electron.

P22.10. An electron (rest mass $= 0.511$ MeV/c^2) is accelerated to a kinetic energy of 10.0 GeV (1 GeV $= 10^9$ eV).
(a) Determine the total energy of the electron.
(b) Find γ.
(c) Find the v/c for this electron.
(d) Find pc for this electron and compare to E.
Reminder: Quantities were given to 3 significant figures.

P22.11. If $v/c = 0.999999$, what is the energy of the electron?

P22.12. The momentum of an electron and photon are both given by $\vec{p}$. Compare the energies and wavelengths of the two particles, that is, find their ratio.

P22.13. The two clocks in a spaceship are synchronized and measure an elapsed time of 1/2 hour. The spaceship moves at $0.8c$. What is the distance covered by the spaceship as measured on earth?

P22.14. An astronaut travels to a distant star and returns to earth. Except for a brief time when she is accelerating or decelerating, the astronaut travels at a speed of 0.995c relative to earth. The star is measured (on earth) to be 40 light years away (40 times the distance light travels in a year). The astronaut spends no time on the distant star.

(a) How long does the trip to the star and back take, as seen by an observer on earth?

(b) How long does the trip take according to the astronaut?

(c) What is the distance traveled, as measured by the astronaut?

(d) If the astronaut and observer are twins, how much younger than the twin on earth will the astronaut be on her return?

Answers to odd-numbered questions

Q1. Yes; $\gamma \approx 1$.

Q3. No, E approaches $m_0 c^2$.

Q5. No; the time in the sun differs, as does the temperature and gravity.

Q7. In A's frame, A is at rest. The time is thus the proper time.

Q9. No. The clock is not at rest relative to the isotope.

Answers to odd-numbered problems

P1. $\Delta mc^2/(mc^2) = mgh/(mc^2) = (9.8 \text{ m/s}^2 \times 4500 \text{ m}/(9 \times 10^{16} \text{ m}^2/\text{s}^2) = 4.9 \times 10^{-13}$.

P3. $\gamma = \frac{1}{\sqrt{1-v^2/c^2}} = 1/0.6 = 1.67$; $t_b = t_A \gamma = 4y/0.6 = 6.67$ y.

P5. $L_{ss} = L/\gamma = 1500 \text{ m} \times 0.6 = 900$ km.

P7. On the moon, $v_{ss} = 0.55c$; $v = \frac{v_1 + v_2}{1 + v_1 v_2/c^2} = \frac{0.55c + 0.55c}{1 + 0.55^2} = 0.85c$.

P9. K.E. $= 3.6 \times 10^7$ eV; $E = $ K.E. $+ m_0 c^2 = 36 + 0.51$ MeV $= 36.5$ MeV.
$E = \gamma m_0 c^2$; $\gamma = E/(m_0 c^2) = 3.65 \times 10^7$ eV$/(0.51 \times 10^6$ eV$) = 71.5$;
$v^2/c^2 = 1 - 1/\gamma^2 = 1 - 1/(71.5)^2 = 0.9998$; $v/c = 0.9999 \approx 1$.
$p = m_0 \gamma v$; $pc = m_0 c^2 \gamma (v/c) = 0.51$ MeV $\times 71.5 \times 1 = 36.5$ MeV.

P11. $\gamma = 1/\sqrt{1-v^2/c^2}$. Since $v \approx c$, let $v/c = 1 - x$; $\gamma \approx 1/\sqrt{1 - 1 + 2x} = 1/\sqrt{2x}$, since $(1-x)^2 \approx 1 - 2x$ when $x \ll 1$; $1 - v^2/c^2 \approx 2 \times 10^{-6}$; $\gamma = 707$ and $E = \gamma m_0 c^2 = 354$ MeV.

P13. $L = v t_e = 0.8c \times 1/0.6 \times 1/2 \text{ hr} = 0.67$ hr.

Chapter 23

Nuclear and Particle Physics

In this chapter we will discuss some aspects of nuclear, particle and astro-physics.

23.1 The Nucleus

In 1897, egged on by the discovery of X-rays by Roentgen, Henry Becquerel decided to look at whether the phosphorescent material (material that glows in the dark after being exposed to visible light) he was examining, emitted X-rays. He exposed his plates to sunlight and then put them on top of photographic plates wrapped in black paper, so that no further light could reach them. He found that the photographic plates were exposed near the phosphorescent material. Further experiments showed that only certain phosphorescent materials, those containing uranium (U) or thorium (Th), produced the effect. One rainy day, Becquerel put the plates in a drawer. Before proceeding, a few days later, he developed the film and found heavy exposure, despite the fact that the material had never been in the sun. He concluded that the sun was unnecessary. Becquerel found that U samples, unlike other materials, retained their penetrating power for weeks. He called the new radiation *natural radioactivity*. It was emitted without any extra energy being provided and left him with a big puzzle.

Marie and Pierre Curie continued studies of radioactivity. They found that a small current was produced when the rays ionized air molecules. The intensity of emission was directly proportional to the mass of metallic elements, e.g. thorium. They discovered a still more radioactive element by separation than Becquerel, which they called polonium, after Marie's birth country, and then a still more active one they called radium. They tried

343

to isolate it. With a 100 kg sample, they got just 0.1 g of RaCl (radium chloride). The atomic mass of Ra was found to be 225, or about 225 times as massive as hydrogen. Radium is more than 10^6 times as active as U.

The rays discovered by Becquerel were studied by others, including Rutherford. He determined that the nucleus of the atom was modified by the emissions. Further experiments showed that the radiation contained three different components, called by the first 3 letters of the Greek alphabet.

By using magnetic and electric fields, it was found that:

(a) alpha particles are the nuclei of $^4_2$He (Z = atomic number = 2, A = mass number = 4),

(b) beta particles were electrons,

(c) gamma rays were high energy photons.

For radioactive decays, the fraction of nuclei that decay ($\Delta N/N$) per unit time is a constant,

$$\Delta N = -\alpha N \Delta t, \tag{23.1}$$

$$N = N_0 e^{-\alpha t}, \tag{23.2}$$

where α is called the decay constant and α^{-1} is the *mean life* of the radioactive nucleus. N_0 is the initial number of decaying nuclei (at $t = 0$). The time for $1/2$ of the initial number of nuclei to decay is called the *half-life*, and is $T_{1/2} = 0.693/\alpha$. Half-lives or mean lives can vary over many orders of magnitude. For instance, the half-life of ^{113}Cd is 9×10^{15}y, whereas that ^{196}Ir is 52 s.

There are many species of nuclei; those with the same *atomic number*, Z, have the same chemical properties. If the neutron number, N, differs, the nuclei are *isotopes*. These nuclei are often made artificially through nuclear reactions. An example of isotopes is $^1_1$H, $^2_1$H, $^3_1$H. The subscript is Z, the number of protons, and the superscript is the mass number. $^1_1$H is ordinary hydrogen, $^2_1$H is sometimes called heavy hydrogen or deuterium, and $^3_1$H is called hydrogen 3.

The proton, the nucleus of the H atom was discovered by bombarding nitrogen with alpha particles. Protons, Z = 1 (Q = Ze = e), mass = 1.673×10^{-23} kg = 980 MeV/c^2, more than 1870 times more massive than the electron, were emitted in the reaction. For many years thereafter, it was thought that the nuclei of atoms are composed of protons (for mass) and electrons (for the electric neutrality of atoms). But there were problems. If electrons are to be contained in a small sphere the size of nuclei, then the uncertainty principle says that the emitted beta rays should have energies of several tens of MeV. However, the energies of beta rays are almost always less than 10 Mev. The solution came in 1932 with the discovery by James Chadwick, who worked with Rutherford, of the neutron. This radiation was even more penetrating than gamma rays. Cavendish bombarded paraffin (carbon and hydrogen) with the radiation and found that protons were emitted forward. From the kinematics of the collision, it was clear that the radiation had a mass about equal to that of the proton. This discovery answered many questions. Nuclei were not made of protons and electrons, but protons and neutrons. The neutron adds mass without charge. Nuclei with the same chemical properties, but different mass numbers are *isotopes* with the same Z, but different N (neutron number).

In a nuclear reaction the charge, and thus Z, is conserved. At low energies the mass number, A, is also conserved. Two examples are $_Z^A nnam$

$$_2^4\text{He} +_7^{14}\text{N} \rightarrow _8^{17}\text{0} +_1^1\text{H}\,, \tag{23.3}$$

$$_2^4\text{He} +_4^9\text{Be} \rightarrow _6^{12}\text{C} +_0^1 n\,, \tag{23.4}$$

where n is a neutron, $_2^4\text{He} = \alpha$. For radioactive decays the initial nucleus is called the *parent* and the resulting nucleus *daughter*,

$$\alpha : _Z^A\text{A} \rightarrow_2^4\text{He} +_{Z-2}^{A-4}\text{A}'\,, \tag{23.5}$$

$$\beta : _Z^A\text{A} \rightarrow_{-1}^0 e +_{Z+1}^A\text{A}'\,, \tag{23.6}$$

$$\gamma : _Z^A\text{A}^* \rightarrow_Z^A\text{A} + \gamma\,, \tag{23.7}$$

where the * indicates a nucleus not in its ground state.

Illustrative Problem 23.1: A nucleus with Z = 23 and A = 50 can emit alpha particles, neutrons, and electrons. What are Z and A for the "daughter" nuclei in each case?

Answer: For α decay, the final nucleus will have Z = 21, A = 46; for neutron decay, the final nucleus will have Z = 23, A = 49; for beta decay, the final nucleus will have Z = 24, A = 50.

23.2 Energetics of Nuclear Reactions

Why are some nuclei stable and others radioactive? This is where the theory of relativity, or rather $E = mc^2$ comes in. If the mass of the nucleus is smaller than that of its decay products, then the nucleus is stable, since it does not have sufficient energy to decay. If the mass is larger than the sum of its products, then the decay occurs and the nucleus is unstable. For an alpha decay, for instance,

$$[m(Z, A) - m(Z - 2, A - 4) - m(2, 4)]c^2 = Q. \qquad (23.8)$$

If $Q \geq 0$, the nucleus is unstable and the excess energy goes into kinetic energy of the products; Q is often called the Q *value* for the reaction. If $Q < 0$, the nucleus is stable.

As an example, consider $^2_1 H \rightarrow p + n$. The masses are given below. The units of mass in nuclear physics are often given in atomic mass units, with $1 \text{ u} = 1.661 \times 10^{-27} \text{ kg} = (1/12) \text{ m}(^{12}_6 C)$.

parent	mass(u)	mass(kg)	mc^2(MeV)
n	1.0087	1.6726×10^{-27}	939.57
p	1.0073	1.6749×10^{-27}	938.27
$^2_1 H$	2.0136	3.3454×10^{-27}	1875.60
$n + p$	2.0160	3.3475×10^{-27}	1877.84

Since the sum of the masses of the neutron and proton is larger than the mass of the deuteron by 2.24 MeV, the deuteron nucleus is stable. This means that a minimum of 2.24 MeV must be supplied to break up deuterium ($^2_1 H$). In the opposite case of $n + p \rightarrow d$, the neutron can be captured at rest and a photon of 2.24 MeV will be emitted in order to balance energy and momentum. The energy of 2.24 MeV is also called the *binding energy* of deuterium, the minimum energy needed to break it up.

The energies in nuclear physics tend to be given in MeV, rather than eV, as in atoms. This is why the radiation is much more penetrating from nuclei than from atoms. Indeed, the radiation from radioactive nuclei can be dangerous to your health, as Marie Curie learned to her chagrin. But radioactive decays and isotopes are used in medicine and in dating ancient objects.

Illustrative Problem 23.2: What is the minimum mass of a neutron to be able to beta-decay to a proton?

Answer: Since the neutron has to emit an electron as well as the proton, the minimum mass is the mass of the proton, 938.27 MeV/c^2, and that of an electron $mc^2 = 0.511$ MeV, for a total of 938.78 MeV/c^2. (The neutron has a mass of 939.57 MeV/c^2. The difference between 939.57 and 938.78 goes to the neutrino and to kinetic energy).

Neutrons and protons tend to be packed tightly in nuclei. This is known from the size of nuclei which grow proportionately to A. Most nuclei are approximately spherical with a radius given by $R = 1.2 \times A^{1/3}$, so that the volume is proportional to A. The energy required to remove a proton or neutron is called the binding energy of that particle. On the average, this energy is about 8 MeV.

Beta decay turns out to be more complex than we have painted it. In a two body decay, e.g., $n \to p + e^-$, momentum conservation tells us that for a decay of a neutron at rest, the momenta of the proton and electron are equal in magnitude, but opposite in direction. The energy released, Q, goes into kinetic energies of the proton and electron. Since $K = p^2/(2m)$, everything is determined and the energies of the electron and daughter nucleus are unique and known. It was found that beta decays did not give rise to electrons of a unique energy, but rather to electrons with an energy distribution, as shown in Fig. 23.1. This was a great puzzle. Niels Bohr was in favor of giving up energy conservation, but Wolfgang Pauli preferred the existence of a neutral particle with a very small mass, dubbed "little neutron" or *neutrino*. This was generally accepted, but it took 30 years to discover the neutrino. By now, physicists have learned to generate neutrino beams and to use them in bombarding nuclear targets. Thus, in

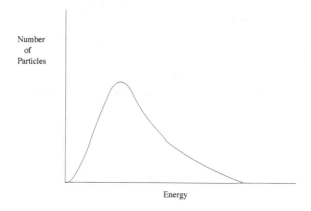

Fig. 23.1 The spectrum of electrons emitted in beta decay.

beta decays, the reaction is really

$$\mathstrut^A_Z A \rightarrow \mathstrut^{A}_{Z+1} A' + e^- + \bar{\nu}_e, \tag{23.9}$$

$$\mathstrut^3_1 H \rightarrow \mathstrut^3_2 He + e^- + \bar{\nu}_e. \tag{23.10}$$

The bar over the ν means that this is an antiparticle. The subscript e has been added because there are 3 species of neutrinos and only one of these is an electron neutrino.

23.3 Fission and Fusion

Enrico Fermi, one of the great physicists of the 20th century, used neutrons to initiate nuclear reactions. The advantage that neutrons have is that they are neutral and are thus not repelled by nuclei, unlike charged protons or nuclei, which tend to be positively charged. The daughter nuclei were always close to those bombarded, showing that an electron, proton, neutron, or alpha particle had been emitted. However, in 1938, Otto Hahn and Fritz Strassman found that the neutron bombardment of U gave daughters that were far away

$$\mathstrut^1_0 n + \mathstrut^{235}_{92} U \rightarrow \mathstrut^{142}_{56} Ba + \mathstrut^{91}_{36} Kr + 3n. \tag{23.11}$$

This was clearly quite different. Lise Meitner and Otto Frisch figured out that the nucleus had become unstable with the neutron bombardment, and broke up into two pieces, see Fig. 23.2. They called the process *fission*.

Niels Bohr and some others realized that a chain reaction could be initiated, since more neutrons were produced than there were initially and the neutrons could initiate further reactions. In order to capture the most neutrons, you want slow ones that stay around the nucleus longer. One complication is that U is mostly ^{238}U in nature, and only 0.7% is ^{235}U. ^{238}U does not fission easily because it tends not to capture neutrons readily. You need to enrich the natural U with ^{235}U. Enrico Fermi did so and built the first nuclear reactor. In order to slow down the neutrons produced in

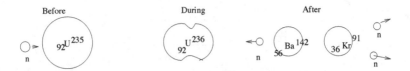

Fig. 23.2   Fission.

the reaction, you need a *moderator* that does so without capturing them. Water is a good moderator. There is a *critical mass* of ^{235}U needed to have a self sustaining reaction. In the process you produce radioactive isotopes that live a very long time and this is one of the difficulties of reactors for energy production. If you can make the fission reaction sufficiently rapidly, then you can make a nuclear explosion with a large amount of available energy.

Fusion is the opposite of fission. Here you create energy by building up nuclei. Light nuclei that have a positive Q value can be chosen. In fact, the sun and other stars are powered by fusion reactions. In the case of the sun hydrogen is turned into $^4_2$He and more hydrogen is produced by the reactions. An example of a fusion reaction is

$$^2_1\text{H} +^3_1\text{H} \rightarrow^4_2\text{He} +^1_0 n\,. \tag{23.12}$$

The reaction produces energy, but to capture the proton you have to overcome the Coulomb repulsion. It is the difficulty of initiating the reaction which prevents us from using this plentiful energy source. In order to initiate it, large lasers are used to concentrate the pellets of hydrogen near the nucleus, and/or large magnetic fields are used to focus them. We have yet to reach the break-even point, where more energy is produced than is being used. The promise is great because we can use sea water for the source of the hydrogen. One of the advantages over fission is that there are no dangerous radioactive substances being produced. In the sun, the probability of initiating the reaction is also small, despite the high temperature, but there are plenty of protons around.

23.4 Particle Physics

Modern physics began with the discovery of the electron. This is one of the *fundamental* particles. Since then, we have met the proton, neutron, alpha, and so forth. By now there are a huge number of new particles that have been discovered and they populate the so-called "particle zoo". Which of these particles, if any, are fundamental, that is particles out of which others can be built? Originally, the Greeks thought atoms were indivisible; this is where the name comes from. In the last several decades it was realized that none of the above particles are fundamental. Atoms are made of electrons and nuclei, nuclei are made of protons and neutrons, and these two objects are made of *quarks*. Are quarks fundamental? They do not have charges that are multiple of e, but the most common two have charges of $2e/3$ and $-e/3$!

Atoms are held together by the Coulomb attraction of electrons to the nucleus. What holds nuclei together? The theory of these so-called "strong" forces is called *quantum chromodynamics*. Unlike the electromagnetic forces, the theory is very nonlinear and must be solved numerically on large computers. At the present time, we know of 4 forces in nature: strong, electromagnetic, weak and gravitational. It is the strong force that holds nuclei together despite the Coulomb repulsion of the protons.

If the strong force is arbitrarily said to have strength 1, then the Coulomb force has a strength of about 10^{-2}, the weak one 10^{-12} and the gravitational one 10^{-38}. The comparison is made at a separation of about 10^{-15}m because the distances over which these forces act, varies. The Coulomb and gravitational forces act over very large distances ($\propto r^{-2}$). The strong force vanishes rapidly for distances large compared to 10^{-15}m; this is called the *range* of the strong force. The weak force acts only over a range of about 10^{-18}m, an incredibly small distance.

The reason we are so familiar with gravity is that the planets and stars are all very massive. In our daily lives, we are not at all aware of the weak force, because of its very short range; however, it is the only force felt by neutrinos.

23.5   Cosmology

There is, of course, a lot of interest in the development of our universe. Despite the large distances involved, the laws of physics apply here as well as to the physics at very short distances.

We believe that the universe began in a huge explosion, called the *big bang*. Ever since then, it has been expanding. This means that every star and galaxy is moving away from us and every other star or galaxy. The distance between galaxies is then approximately $d = vt$, where v is the velocity of separation and t is the time from the beginning of the universe. An apt analogy is to a balloon that is being blown up. Any spot on the surface of the balloon gets further and further away from another spot. The expansion of space was first noted by Edwin Hubble in the 1920's. He noted that the velocity of separation increases with distance, $v = Hd$, where H is called the Hubble constant. H has the dimension of inverse time. The expansion is responsible for the *red shift* of observed spectra, for instance, hydrogen from stars. For a constant expansion rate, we can estimate the age of the universe from Hubble's "law" as $t = 1/H$, about 13 billion (13×10^9) y. It was discovered several years ago, that the expansion has been increasing in rate.

Cosmologists classify matter as *baryonic* (from the Greek "barys", meaning heavy), or ordinary matter, and non-baryonic, including neutrinos. It is also now known that baryonic matter only makes up about 5% of the matter in the universe. About 30% is "dark matter" and almost 70% is "dark energy." We do not know what constitutes dark matter, but it is known to have gravitational effects; we know even less about dark energy at this time. The hunt is on to discover these two strange energies that fill the universe.

Questions

Q23.1. In bombarding $^{14}_{7}$N with alpha particles, protons are produced. What is the nucleus produced with the emission of a proton?

Q23.2. If a neutron is captured by $^{238}_{92}$U, what is the resulting nucleus? Also provide its Z and A numbers.

Q23.3. Can two atoms of the same chemical element have different masses? Explain.

Q23.4. In a time equal to 5 half lives of a radioactive isotopes, is all of the isotope gone? Explain.

Q23.5. Can the bombardment of an element (e.g., U) with neutrons lead to heavier elements? Can you do so if you use magnesium ions (a magnesium atom with one or more electrons removed) instead of neutrons? Explain.

Q23.6. $^{131}_{53}$I beta decays. What are the product elements?

Q23.7. The nucleus $^{8}_{4}$Be is unstable and emits an alpha particle. What is the daughter nucleus?

Q23.8. Are quarks contained in electrons?

Q23.9. How would you make up the proton and neutron out of the quarks cited in the text?

Q23.10. In a fission reaction of $^{235}_{92}$U with a slow neutron, $^{140}_{54}$Xe is produced together with 2 neutrons. What is the other nucleus?

Q23.11. The antiparticle to an electron is called a positron. Some radioactive nuclei emit positrons in their decays. What is the equation for the decay of ^A_ZA to a positron?

Problems

P23.1. In the reaction $^4_2$He $+^9_4$Be $\rightarrow^{12}_6$ C $+ n$, the masses are as follows: Be: 9.0122 u, He: 4.0026 u, C: 12.0000 u (by definition), and n: 1.00867 u. Is this reaction exothermic or endothermic? If energy is needed, how much energy in MeV. If energy is released, how much energy, in MeV, is available for kinetic energy of the two nuclei?

P23.2. In the fusion reaction $^2_1$H $+^3_1$ H $\rightarrow^4_2$ He $+ n$, how much energy is released? The masses are ^{2}H: 2.014102 u, ^{3}H: 3.016050 u, ^{4}He: 4.002603 u, n: 1.008665 u. Give the energy in MeV.

P23.3. How many half-lives have gone by when you are left with 1/32 of a radioactive substance?

P23.4. A pion is a particle of mass 140 MeV/c^2.
(a) A neutral pion decays into two gammas. What are their energies and momenta in MeV and MeV/c? (Note that matter has been transformed into electromagnetic radiation).
(b) A negatively charged pion beta decays into a negative muon (mass = 105 MeV/c^2), $\pi^- \rightarrow \mu^- \bar{\nu}\mu$. What is the energy of the muon?

P23.5. The binding energy of a proton in a nucleus is the energy needed to remove it. The mass of $^{121}_{51}$Sb is 120.90382 u; that of $^{120}_{50}$Sn is 119.90220 u. Determine the binding energy of the proton ($m = 1.00783$ u) in Sb.

P23.6. You can create matter from energy, according to Einstein. If we have two gamma rays collide, you can produce a proton and its antiparticle (opposite charge and same mass as proton). What energy is required in MeV?

Answers to odd-numbered questions

Q1. $^{14}_{7}$N $+^{4}_{2}$ He $=^{1}_{1}$ H+? The question mark has to be $_{8}$O^{17}.

Q3. Yes, e.g., isotopes.

Q5. Yes, if the neutron is captured; e.g., $n + U \rightarrow U + \gamma$. Yes, if a lighter ion is emitted.

Q7. $^{4}_{2}$He.

Q9. uud and udd.

Q11. $^{A}_{Z}$A $\rightarrow^{A}_{Z-1}$ A$'$ $+^{0}_{1}$ e^{+}.

Answers to odd-numbered problems

P1. In the initial state the masses are $9.0122 + 4.0026 = 13.01480$ u. In the final state, the masses are $12.000 + 1.00867 = 13.00867$. The reaction is exothermic. The difference is 0.00613 u $= 0.00613 \times 1.66 \times 10^{-27}$ kg $\times 9 \times 10^{16}$ m^{2}/s^{2}/$(1.6 \times 10^{-19}$ J/eV$) = 5.7$ MeV is available for kinetic energy.

P3. $(\frac{1}{2})^{5} = 1/32$; 5 half-lives.

P5. $120.91003 - 120.90382 = 0.00621$ u since $119.90220 + 1.00783 = 120.91003$ u. The binding energy of the proton is therefore 931 MeV $\times 0.00621 = 5.78$ MeV.

Index of Famous Scientists

Subject Index